Mathematics Teaching and Learning in K–12

Mathematics Teaching and Learning in K–12

Equity and Professional Development

Edited by
Mary Q. Foote

MATHEMATICS TEACHING AND LEARNING IN K-12
Copyright © Mary Q. Foote, 2010.
Softcover reprint of the hardcover 1st edition 2010 978-0-230-62239-5

All rights reserved.

First published in 2010 by
PALGRAVE MACMILLAN®
in the United States—a division of St. Martin's Press LLC,
175 Fifth Avenue, New York, NY 10010.

Where this book is distributed in the UK, Europe and the rest of the world, this is by Palgrave Macmillan, a division of Macmillan Publishers Limited, registered in England, company number 785998, of Houndmills, Basingstoke, Hampshire RG21 6XS.

Palgrave Macmillan is the global academic imprint of the above companies and has companies and representatives throughout the world.

Palgrave® and Macmillan® are registered trademarks in the United States, the United Kingdom, Europe and other countries.

ISBN 978-1-349-38413-6   ISBN 978-0-230-10988-9 (eBook)
DOI 10.1057/9780230109889

Library of Congress Cataloging-in-Publication Data

    Mathematics teaching and learning in K-12 : equity and professional development / edited by Mary Q. Foote.
      p. cm.
    Includes bibliographical references.

    1. Mathematics—Study and teaching—United States. 2. Multicultural education—United States. 3. Educational equalization—United States. I. Foote, Mary Q.

QA13.M369 2010
510.71'073—dc22   2009049124

A catalogue record of the book is available from the British Library.

Design by Newgen Imaging Systems (P) Ltd., Chennai, India.

First edition: July 2010

This book is dedicated to the scholars who were pioneers in bringing an equity lens to the field of mathematics education. They have played a significant role for me and for the other authors of this volume. It is upon their work that ours is built, as we acknowledge in our numerous citations of their valuable scholarship.

CONTENTS

Preface ix

Acknowledgments xi

Introduction 1
Mary Q. Foote

Part I Who is Participating in the Professional Development?: Examining the Trajectories of Teacher Engagement

1 Bilingual Elementary Teachers' Reflections on Using Students' Native Language and Culture to Teach Mathematics 7
Sylvia Celedón-Pattichis, Sandra I. Musanti, and Mary E. Marshall

2 Centering the Teaching of Mathematics on Students: Equity Pedagogy in Action 25
Laurie H. Rubel

3 The Power of One: Teachers Examine Their Mathematics Teaching Practice by Studying a Single Child 41
Mary Q. Foote

4 Pursuing "Diversity" as an Issue of Teaching Practice in Mathematics Teacher Professional Development 59
Ann Ryu Edwards

5 Teacher Positioning and Equitable Mathematics Pedagogy 77
Anita A. Wager

6 Math Is More Than Numbers: Forging Connections among Equity, Teacher Participation, and Professional Development 93
Carolee Koehn

7 Using Artifacts to Engage Teachers in Equity-based Professional Development: The Journey of One Teacher 111
Kristine M. Ho

Commentary: Part I 129
Kyndall Brown and Megan Franke

Part II What Tools Have Proved Effective?: Examining Effective Resources and Vehicles for Addressing the Needs of Teachers and Their Students

8 Building Community and Relationships that Support Critical Conversations on Race: The Case of Cognitively Guided Instruction 137
Dan Battey and Angela Chan

9 Status and Competence as Entry Points into Discussions of Equity in Mathematics Classrooms 151
Victoria M. Hand, Jessica Quindel, and Indigo Esmonde

10 Creating "Constructive Opportunities": A "How" to Embracing Students' Mathematical Conceptions 167
Vanessa R. Pitts Bannister, Gina J. Mariano, and Carla D. Hall

11 Using Lesson Study as a Means to Support Teachers in Learning to Teach Mathematics for Social Justice 181
Tonya Gau Bartell

12 Keeping the Mathematics on the Table in Urban Mathematics Professional Development: A Model that Integrates Dispositions toward Students 199
Joi Spencer, Jaime Park, and Rossella Santagata

Commentary: Part II 219
Megan Franke and Kyndall Brown

References 225

Notes on Contributors 239

PREFACE

Contributions for this volume were initially solicited from among current and former fellows of the Diversity in Mathematics Education, Center for Teaching and Learning (DiME/CLT), funded by the National Science Foundation. One goal of the center was to develop researchers who will focus their research agendas on the important problem of addressing equity and diversity concerns within mathematics education. Through DiME many of the authors of chapters in this volume came together to be part of a project whose goals also included a focus on the professional development of teachers of mathematics, taking into consideration the issues of equity and diversity that affect their teaching.

DiME has a history of collaborative work and has already engaged in important scholarly activities. After two years of collaboration dedicated to studying issues framed by the question of why particular groups of students (i.e., poor students, students of color, English learners) fail in school mathematics in comparison with their White (and sometimes Asian) peers, DiME presented a symposium at an Annual Meeting of the American Educational Research Association (DiME, 2005). This was followed by the writing of a chapter in the recently published *Second Handbook of Research on Mathematics Teaching and Learning* (DiME, 2007).

Many former DiME fellows are now professors at a variety of universities across the United States, and in many cases, they have teamed with other researchers, practitioners, and graduate students of their own to produce chapters in this volume. In addition, they have forged links with other researchers at their new universities or whom they have met while working on other projects. Some of those researchers have also contributed to this book. The volume, thus, represents the collective efforts of a group of people who began working together (and have begun to extend that working group) to frame a common problem and to engage in research to suggest solutions to that problem.

Mary Q. Foote
New York City

ACKNOWLEDGMENTS

Many of the contributors to this volume owe a particular debt of gratitude to the National Science Foundation through its funding of the Diversity in Mathematics Education, Center for Learning and Teaching (DiME/CLT). More specifically, the research reported on in the chapters by Bannister, Mariano, and Hall; Bartell; Battey and Chan; Edwards; Foote; Ho; Koehn; and Wager and the portion of Chapter 9 reporting on the research of Hand is based upon work supported by the National Science Foundation under Grant ESI-0119732 awarded to the DiME/CLT. The research reported on in the chapter by Celedón-Pattichis, Musanti, and Marshall is based upon work supported by the National Science Foundation under Grant ESI-0424983 awarded to the Center for the Mathematics Education of Latinas/os, Center for Learning and Teaching (CEMELA/CLT). The research reported on in the chapter by Rubel is based upon work supported by the Knowles Science Teaching Foundation and the National Science Foundation under Grants 0742614, 0119732, and 0333753. The research reported on in the chapter by Spencer, Park, and Santagata is based upon work supported by the Institute of Educational Sciences (IES, Teacher Quality Program) under Grant R305M030154. Any opinions, findings, and conclusions or recommendations expressed in this volume are those of the authors and do not necessarily reflect the views of the funding organizations.

In addition to the National Science Foundation and the other funding organizations, there are others I would like to acknowledge. The three principal investigators of the DiME/CLT, Tom Carpenter (my dissertation advisor, to whom I am especially grateful), Megan Franke, and Alan Schoenfeld, have nurtured and supported many of us in our academic careers; to them all I am very grateful. More generally, I would like to thank the members of the DiME collaborative from whom I draw much strength. I am particularly grateful to Meg Meyer, a steady and supportive voice on the other end of the phone, always there to hear about the successes and frustrations involved in editing this volume. Thanks to Julia Cohen, my first Palgrave Macmillan editor, for offering me the opportunity to compile this volume and for guiding me in the early stages of the project and to Samantha Hasey, my subsequent Palgrave Macmillan editor, for her patience and understanding with my many questions as I worked on the final editing. I would like to extend my appreciation to my colleagues at Queens College, CUNY, and my close friends and family, who have been so supportive during the time I worked on this volume. Finally, I would like to thank my children, Kate and Joe, who bring me so much joy and from whom, throughout many years, I have learned so much.

Introduction

Mary Q. Foote

Attending to issues of equity and diversity within the field of Mathematics Education is a relatively new occurrence. Often in the past, research in mathematics education was looked at as culture free, while issues of equity and diversity were seen as applicable uniformly across any content. More recently, however, a growing group of mathematics educators have begun to bring together the two fields. This is at least in part born of an understanding that all teaching and learning happens within a culturally situated context from which it cannot be divorced and that a deeper understanding of each enriches both.

It is not only within the mathematics education research community that mathematics has been considered culture free, but it is also within preK-12 school settings that this has occurred. The continuing gap in achievement between traditionally underserved students (students of color, English learners, and poor children) and their middle-class White peers, however, has provoked questions of the effectiveness of current mathematics teaching practices for meeting the needs of these students. Professional development of teachers of mathematics has been identified as a way to address not only teaching practices but the beliefs and orientations that teachers have that may interfere with their being effective teachers for students unlike themselves (Howard, 1999; Nieto, 2004). Mathematics educators who sometimes plan and facilitate as well as study professional development for teachers of mathematics have begun to examine the effects of merging issues in mathematics teaching and learning with issues of equity and diversity within professional development efforts. This book reports on a number of these recent studies.

The chapters in this book are divided into two parts, one focusing on teachers and their trajectories of growth and development, and one on the particular vehicles for engaging with the professional development content (the blending of equity and mathematics). The distance between research and practice in education is something that is often remarked on. This book is intended to speak to both educational researchers (in mathematics education as well as in equity studies) and teacher practitioners, to support a dialogue between them about the important issue of developing effective teaching practices in mathematics that encourage and acknowledge the participation and achievement of all students, not just those who have historically been successful. The majority of university-based authors of these chapters came into educational research from classroom positions and so have an understanding of the importance of attending to both research issues and practical considerations in teaching children in preK–12 schools. Some of these researchers are teaming with K–12 teachers in authoring these chapters, providing an even more intense focus on real students in real classrooms and the consequences of professional development efforts for teaching them and for their learning. To emphasize even further the importance of these studies to both the research community and the K–12 teaching community, each part of the book will conclude with commentaries on the chapters within that section. These commentaries are written by two scholars: one, Megan Franke, whose work is centrally involved with research in teacher professional development in mathematics education, will bring a research perspective to the commentaries; the other, Kyndall Brown, who has worked as a professional developer of mathematics teachers for many years, will bring a solid practice-based perspective to the commentaries. The dual focus of these commentaries is meant to underscore the fact that educational research must have strong links to practice.

In Part I, the trajectories of participation and teacher identity development as they evolve during the professional development are examined. The importance of how teaching and learning play out in local contexts makes the question of who is involved in professional development a particularly important one. In some cases, such as that presented by Wager, the trajectory is one based on individual histories. In others, such as that discussed by Edwards, it is the group history that influences the trajectory. Celedón-Pattichis, Musanti, and Marshall report on a study in which researchers worked directly with teachers to plan, reflect on, and design further mathematics instruction to address the specific needs of bilingual students. Rubel describes a professional development program aimed at supporting teachers in "centering" mathematics instruction of urban youth on the students themselves, including their mathematical thinking and their lived experiences. Foote discusses a professional development in which the teacher's gaze was directed outside of the classroom in order to discover competencies the child had that could be build on in the mathematics classroom. Edwards compares the trajectories of two groups of teachers, working in a district newly committed to teaching Algebra to all middle school students, as they grapple with notions of diversity and equity and their implications for their teaching practice. Wager examines the trajectories of progress of three teachers engaged in a professional development focused on examining and enacting an equitable mathematics pedagogy. Koehn describes how groups of teachers working together during the professional development, while addressing both intellectual and emotional engagement, were able to increase their ability to approach nonroutine problems. And Ho examines how the use of artifacts of practice (student work, teacher autobiographies, and lesson plans) helps expose interactions between race, equity, and mathematics learning.

Structures that have been shown to be effective in engaging teachers in addressing the needs of traditionally underserved students are the focus of the chapters in Part II. The blending of equity issues and issues of mathematics teaching and learning in professional development can be challenging, and it is important to identify and examine structures that can be enlisted to support this effort. Battey and Chan discuss a study that used engagement with student thinking in mathematics to challenge deficit stories teachers had of some students in their classrooms. Hand, Quindel, and Esmonde examine how the very nature of the classroom group work can be used as a lens into differences in student participation that may be the result of experiences within and outside of the classroom. Pitts Bannister, Mariano, and Hall focus on supporting teachers in admitting into the classroom the diverse mathematical notions that students have developed in their lived experiences, as an initial step toward equitable teaching. Bartell describes using lesson study as a vehicle to support teachers in incorporating both mathematics and social justice goals into their lessons. Lastly, Spencer, Park, and Santagata bring a different perspective. They examine the inadequacies of a particular professional development structure to improve practice and advocate a need to look at teachers' dispositions toward students as that can influence, oftentimes for worse, the orientation those teachers have to teaching conceptually challenging mathematics.

Together these chapters present a picture of possibility for engaging teachers in the hard work of examining and changing their practice. These professional development contexts support teachers in developing deeper content and pedagogical content knowledge. They also encourage teachers to question the beliefs, assumptions, orientations, and dispositions they bring to the teaching of mathematics, particularly to students who the educational system in the United States has not served well.

PART I

Who is Participating in the Professional Development?: Examining the Trajectories of Teacher Engagement

CHAPTER 1

Bilingual Elementary Teachers' Reflections on Using Students' Native Language and Culture to Teach Mathematics

SYLVIA CELEDÓN-PATTICHIS, SANDRA I. MUSANTI, AND MARY E. MARSHALL

In mathematics education, equity means that all students regardless of their language, ethnicity, gender, and socioeconomic status have the right to engage in high-quality classroom experiences that support meaningful mathematics learning (Secada, 1989, 1992, 1995). This is clearly stated in the National Council of Teachers of Mathematics (NCTM) Position Statement on Teaching Mathematics to English Language Learners and the Equity Principle of the *Principles and Standards for School Mathematics* (NCTM, 2000, 2008). Considering the fact that Latina/o students are the fastest-growing group in public schools, that nearly half (45 percent) are English language learners (Kohler & Lazarín, 2007), and that there exist large opportunity learning gaps for Latinas/os (Flores, 2007), there is a sense of urgency to better prepare teachers to address the needs of all students, particularly Latina/o students. Unfortunately, there are few certified teachers who are adequately prepared to teach culturally and linguistically diverse students (Téllez, 2004/2005). Professional development opportunities for bilingual teachers are usually centered on discussions of language and literacy, but rarely do they combine subject-matter knowledge, such as mathematics (see Musanti, Celedón-Pattichis, & Marshall, 2009).

In this chapter we present the findings of a four-year longitudinal study in mathematics professional development. The purpose of this qualitative study was to strengthen the practices of K-2 bilingual teachers' professional development experiences in the subject area of mathematics. Specifically, this study explores the following research questions: (a) How can classroom-based professional development further bilingual teachers' understanding of K-2 Latina/o students' problem-solving strategies and mathematical thinking in their first language, which in this case is Spanish, and (b) What were teachers' reflections upon their experience with this classroom-based approach to promote mathematical learning? Specifically, we focus on bilingual teachers' reflections on their use of students' native language and culture to teach mathematics. In the sections that follow we present the theoretical perspectives that inform this work and the methodology. We end with a discussion of the findings, implications, and future directions of our work.

Literature Review

We draw from three bodies of literature to situate our work. One involves Cognitively Guided Instruction (CGI) (Carpenter, Fennema, Franke, Levi, & Empson, 1999), a framework that focuses on understanding students' strategies for solving context-rich word problems. The second includes literature that explores teacher learning through professional communities that engage teachers in ongoing reflection on the everyday task of teaching (Crockett, 2002; Little, 2005; Wenger, 1998). We discuss how professional development is designed so that teachers learn in collaboration with researchers and vice versa, how we provide opportunities to reflect on teachers' practice through debriefing sessions by analyzing student work, and how we collegially design teaching approaches that respond to students' needs. The third piece of literature consists of Cummins's (1986, 2001) work on supporting the teaching of mathematics through the students' native language.

Cognitively Guided Instruction

The CGI framework is based on specific types of problems and the strategies that students use to solve them. The basic premise of CGI is that students bring to school intuitive knowledge about problem solving that can help them bridge their informal numeric understanding with formal mathematics. CGI problem types target the concepts children are learning in school, such as addition and subtraction, so that children's out of school experiences become the foundation for their formal mathematical learning.

The research project incorporated professional development opportunities for teachers who were willing to learn about mathematical reasoning, problem solving, and language and culture issues in the mathematics learning of Latina/o students. Following the CGI framework (Carpenter et al., 1999), a central premise that guided our approach to professional development was to foster teachers' understanding of the relevance of problem solving in mathematics education. Underlying this premise is the centrality of promoting "learning for understanding" and the need to form teachers who know how to help students:

> (a) connect knowledge they are learning to what they already know, (b) construct a coherent structure for the knowledge they are acquiring rather than learning a collection of isolated bits of information and disconnected skills, (c) engage in inquiry and problem solving, and (d) take responsibility for validating their ideas and procedures. (Carpenter et al., 2004, p. 5)

Reflecting Within Communities of Practice

Following Wenger (1998), a social conceptualization of learning involves understanding it as taking place within communities of practice in which meaning construction happens as learning impacts our identity, creating "personal histories of becoming in the context of our communities" (p 5). For individuals, then, learning entails engaging in shared enterprises and participating in the collective negotiation of meaning through practice. Therefore, constituting a community of practice involves mutual engagement, a joint enterprise, and a shared repertoire of resources for negotiating meaning: "Such communities hold the key to real transformation—the kind that has real effects on people's lives" (p. 85).

Research has shown that teachers' feelings and beliefs about mathematics, their conceptions of mathematics teaching and learning, and their experiences as learners carry on into their instructional practice (Aguirre, Celedón-Pattichis, Musanti, & Anhalt, 2009; Phillip, 2007; Thompson, 1992). To deepen our understanding of the connection between teachers' affects and conceptions of mathematics, and their teaching approach, research on teachers' beliefs cannot focus solely on verbal data but must also include direct examination of teachers' instructional practice (Thompson, 1992). Moreover, Thompson argues that to understand how teachers' conceptions change as a consequence of participating in professional development programs "it should seem necessary to study individual teachers in depth and to provide detailed analyses of their processes" (p. 40). Our approach to professional development concurs with Philipp's (2007) stance that

> beliefs might be thought of as lenses through which one looks when interpreting the world, and affect might be thought of as a disposition or tendency one takes toward some aspect of his or her world; as such, the beliefs and affect one holds surely affect the way one interacts with his or her world. (p. 258)

Another premise of our approach was that "professional development opportunities should engage teachers in what teachers do" (Crockett, 2002). Therefore, teachers were offered varied opportunities to reflect on their practice, discuss activities and their daily work, design lessons appropriate for students' needs and grade level, and reflect on student work. Recently, researchers have promoted the use of student work as a tool to engage teachers in reflection on students' learning and thinking (Ball & Cohen, 1999; Kazemi & Franke, 2004; Little, 2005). Student work was an important catalyst for reflection on mathematical problem solving and took place in different manners. For instance, teachers discussed video clips of young Latina/o students working on challenging problems and commented on how students used different strategies to solve problems, what students struggled with, and how the teacher facilitated mathematical discussions. In addition, teachers had the opportunity, typically during in-class support, to reflect on different pieces of student work produced in their own classrooms.

Using Native Language and Funds of Knowledge in Mathematics Teaching

We view the role of the native language as a human right (Skutnabb-Kangas, 2000) and as a resource (Moschkovich, 2007) in the mathematics classroom. The important role that the native language plays in teaching mathematics has been documented in past research (Celedón-Pattichis, 2008; Jensen, 2007; Khisty, 1995; Moschkovich, 2007; Secada & De La Cruz, 1996; Turner, Celedón-Pattichis, & Marshall, 2008). Research has shown that making mathematics concepts accessible to students in their native language helps them to transfer the same concepts from the first language to the second language (Cummins, 1986, 2001). For example, when a student learns to multiply in the first language that concept does not have to be relearned in the second language. It is the new label of the word (multiplication) that students need to learn. However, the academic language that students need to perform cognitively demanding tasks in mathematics can take from four to seven years or more to develop in the second language (Cummins, 1986; Thomas & Collier, 2002). Simultaneously, learning a second language and mathematics is a cognitively demanding task that can slow down the process of mathematics learning. On the other hand, learning the mathematics concepts in the first language, the language of proficiency, allows students to make sense of complex mathematical ideas without the burden of also learning a new language. Later, once students have learned the mathematics in their first language, they need to assign new labels to their already formed understandings. Secada and de la Cruz (1996) argue that in order to raise the unacceptable mathematics scores of Latinas/os, students must learn mathematics with understanding (Hiebert & Carpenter, 1992). Making the native language accessible to students is a way to remove the ambiguity that is sometimes present in the context of problem solving. Thus, instruction in the native language is important to access the language needed for mathematics problem solving while students learn a second language.

In addition to language, we view the students and their communities' funds of knowledge as intellectual resources (Civil & Andrade, 2002; González, Moll, & Amanti, 2005) that inform teaching and learning in the mathematics classroom. Funds of knowledge include the knowledge, skills, and cultural practices of students' households that are valued and encouraged in the mathematics classrooms. Teachers can help students learn with understanding by knowing the communities that students come from and the mathematical practices in which their families engage. For example, Turner and her colleagues (2008) found that storytelling practices used with kindergarten Latina/o children engaged students in meaningful mathematics problem solving. Teachers framed problem solving around telling and sharing stories. This practice drew upon familiar ways of talking and negotiating meaning (Delgado-Gaitan, 1987; Villenas & Moreno, 2001). The context of these stories reflected similar storytelling practices of students' families and allowed teachers ways to scaffold students' mathematical thinking.

Methodology

Participants

Three Latina elementary bilingual certified teachers participated in this study[1]. Their teaching experience ranged from three to 20 years. For the purpose of this study, we will consider novice teachers as those who had five or less years of experience and experienced teachers as those with more than five years. Table 1.1 shows teachers' placement, experience, type of endorsement, and past professional development. Although we worked with a unique set of teachers regarding their knowledge of bilingual education, none of them had had experience with CGI prior to the initiation of this study.

We selected these participants because they had participated in one or more of the three options for professional development—summer institute (SI), in-class support, or workshops; they were willing to integrate problem-solving lessons into their reform-oriented curriculum; and they were teaching in bilingual classroom settings with a high percentage of Latina/o students. The authors provided the professional development activities and were able to collect information from each teacher throughout the duration of this initiative.

Participants were teaching at an elementary school located in a major city in the southwestern United States. The school had a predominantly Latina/o student population (over 85 percent) characterized as low income, with a high percentage of students receiving free or reduced price meals (more than 90%). Since most of the students' native language was Spanish, the school had a 90:10 bilingual program in which Spanish was used 90 percent and English 10 percent of the time in kindergarten, 80 percent in first grade, gradually increasing the use of English to 50 percent in fourth and fifth grades (Lindholm-Leary, 2001). The three teachers spoke Spanish as their first language and were teaching in bilingual classrooms in which mathematics was taught in Spanish. One English as a second language class was offered to students in the afternoon to teach basic academic vocabulary in English.

Table 1.1 Participating teachers' information

Name	Grade Level	Teaching Experience	Bilingual or ESL Endorsement	Summer Institute	In-class support	Workshops or Teacher Study Group
Elba	K	Experienced	Bilingual	Two	Yes	Yes
Karmen	K	Novice	Bilingual	One	Yes	Yes
Norma	1st	Experienced	Bilingual	One	Yes	Yes

Professional Development Design

We were interested in the impact of a series of professional development opportunities on the instructional practices of early elementary bilingual mathematics teachers. Over 30 teachers were engaged in the professional development, which included three components: (a) two SIs during 2005 and 2006, (b) workshops, and (c) sustained in-class support. Teachers were selected on the basis of their interest in learning about the integration of language, culture, and mathematics. Participating teachers came from a variety of schools and were either novice or experienced.

Summer institutes. The SIs involved two intensive weeks of class during the month of June. Participating teachers received academic credit for attending and completing the course requirements. The purpose of the first SI was to deepen teachers' understanding of mathematical problem solving and issues of language and culture in mathematics teaching and learning. The first SI included a three-hour graduate level course focused on integrating language, culture, and mathematics and was taught by the first author. The goal of this course included having teachers understand the nuances of mathematics language and moving teachers beyond a focus on teaching vocabulary.

Although the three teachers were very knowledgeable about the importance of funds of knowledge in teaching English learners and about theories of first and second language development, during these SIs all of them expressed the need to connect that knowledge specifically to mathematics teaching and learning (see teacher's questionnaire later in this chapter). In addition, CGI training was incorporated into the SIs. The second SI consisted of a three-hour graduate level course on teaching mathematics with a focus on facilitating discussions for all students, especially second language learners. An average of 30 teachers participated in each institute. All participants attended one or both of the SIs.

Workshops. The workshops took place six times during 2007–2008 and involved teachers from one elementary school in the same school district. The workshops, conducted by the authors, focused on learning about CGI and supporting teachers who were interested in implementing it in their classrooms. The group met for two hours. The three teachers participated in the workshops during 2007–2008.

In-class support. In-class support involved researchers' frequent visits to the teachers' class to observe mathematics lessons, participate in or model CGI problem-solving lessons, discuss different ways to implement problem-solving activities, provide resources to supplement mathematics curriculum, and offer time for debriefing conversations to discuss classroom events related to mathematics instruction. Providing teachers with the opportunity to collaborate with researchers in the classroom was central to our belief that teachers should be afforded opportunities to learn from and within the teaching context (Ball & Cohen, 1999). The three teachers received in-class support from the authors during a period that varied from one to three years.

Data Collection

For the longitudinal project, multiple data were collected in order to support data analysis with regard to our different research questions and as a means to assure data triangulation. For this study, we drew from two main sources of data: interviews with each of the participating teachers and videotaped classroom sessions from three teachers (Karmen, Elba, and Norma). In addition, researchers collected field notes from our visits to teachers' classrooms and kept a log or journal with notes from the meetings with each teacher.

Interviews. Two interviews were conducted with the three teachers during the school year. Each lasted approximately 45 minutes. The interviews explored teachers' reflections of mathematics curriculum and teaching, the nature of mathematics, curriculum integration, the impact of culture and language on mathematics teaching, and teachers' knowledge of students. Interviews were audiotaped and later transcribed.

Classroom observations. The three teachers were observed on a regular basis while implementing CGI problem-solving lessons and as part of the in-classroom support provided by researchers. Elba, Karmen, and Norma, were visited on a weekly basis. Field notes were taken during these observations that typically lasted between 30 and 45 minutes. The nature of the field notes typically included the type of word problems presented, how the teacher used scaffolding to present mathematics language even when instruction was in Spanish, and how the teacher drew from students' cultural practices to connect to mathematics teaching and learning. We videotaped 34 lessons in Elba's class, 25 in Norma's class, and 25 in Karmen's class.

Teachers' questionnaire. Teachers who participated in one or both of the SIs voluntarily answered a cultural identities and mathematical teaching (CIMATH) questionnaire (Aguirre, Celedón-Pattichis, Musanti, Anhalt, 2009). We used this questionnaire to obtain data on teachers' reflections prior to initiating professional development activities and one year after providing various types of support. The questionnaire consisted of six open-ended questions that asked teachers to explain and give examples of (a) the reasons why students do well or have difficulty in learning mathematics, (b) teaching strategies that help all students learn mathematics, (c) strategies to help English language learners learn mathematics, and (d) ways in which students' and teachers' cultural identities inform the teachers' vision for teaching mathematics.

Data Analysis

To explore teachers' reflections on the use of native language and culture to teach mathematics, we applied a constant comparative method (Lincoln & Guba, 1985; Strauss & Corbin, 1998) of analysis to our data transcripts, including interviews and field notes from classroom observations. This approach allowed us to explore teachers' responses and actions in search for patterns and contradictions. Initially, the research team individually open coded (Strauss & Corbin, 1998) the interview transcripts looking for themes in teachers' responses that illustrated their reflections on native language use, culture, impact of professional development on their own teaching and on students, the mathematical content they valued, the place of problem solving in their mathematics instruction, and the sources they considered reliable as they made decisions about curriculum and appropriate teaching practices. Later, the research team met extensively to discuss the conjectures raised from the initial data coding and to collectively code the transcripts, define codes, and integrate them into more inclusive categories. Constructing conceptual categories was important because they allowed us to "capture some recurring patterns that cut across...the data" (Merriam, 1998, p. 179). Data were managed using a computer-based qualitative research tool, TAMS© Analyzer. This tool allowed us to create and define codes and later refine them into more inclusive categories. Video recordings, transcripts, field notes, and teachers' questionnaire were important elements of triangulation.

Trustworthiness

According to Erlandson and colleagues (1993), one of the ways to establish trustworthiness is through long-term engagement. We began our work with these teachers four years ago upon beginning the

NSF-funded project, Center for the Mathematics Education of Latinas/os (CEMELA). During this four-year period, we have developed personal relationships with these teachers and have come to know the successes and challenges encountered by each in her classroom.

In the next section, we discuss what we learned from these bilingual teachers as they reflected on their teaching of mathematics. They often discussed the importance of building academic language and how they adapted the curriculum as they considered students' culture.

Findings

Building Academic Language

Supporting Latina/o students' academic language development, even while teaching in students' native language, is challenging but crucial to their future academic success. Research has reported that Latina/o students' low graduation rate (45 percent of Mexican American students do not complete high school) is due not only to their lack of English proficiency but also to the lack of opportunities they are afforded to develop a strong academic discourse in Spanish (López-Bonilla, 2002). Undoubtedly, minority students who have the opportunity to learn in the context of a bilingual program transfer the skills learned in their first language to the second language (Cummins, 1986, 2001).

Norma, Elba, and Karmen reflected about the difference between everyday language and academic language (Bielenberg & Wong Filmore, 2004/2005; López-Bonilla, 2002; Valdés, 2004). They often expressed concerns about the academic and testing demands the students were exposed to and the need to generate instructional moves that would improve students' academic language. Many of our conversations focused on how to build on the differences between academic and everyday language, and since the beginning of our collaboration, these teachers have been grappling with understanding the interconnection of both in the context of instruction:

> Well, I look at the academic language, native language—Spanish—and the everyday language. And it is not the same. And the majority of the kids that I have here... their level, it's not limited language, but it's daily, everyday common language. (Norma's Interview, November 2006)

The notion of academic language should be defined not only in terms of the specific structures and a specialized lexicon but with reference to the sociocultural elements that integrate socially accepted discourse (López-Bonilla, 2002). Initially, Norma's and Elba's conception of academic language seemed to be restricted to the introduction of specialized vocabulary.

> If you are not used to explaining your reasoning, in Spanish, or whatever language you talk, if you don't know the vocabulary, you are not going to be able to explain.... When we look at specific words that they use, specific vocabulary that we use in math... it's not a vocabulary that you use everyday, so it has to be taught and practiced. (Norma's Interview, November 2006)

★★★

> And that's another difficult thing.... If they don't have the vocabulary, they cannot explain their reasoning. (Elba's Interview, Spring 2006)

However, as our work progressed and we observed the teachers' pedagogy and interaction with students, we conjectured that their emphasis on improving students' vocabulary did not disregard other linguistic and cultural elements and students' meaning construction in the context of mathematics learning. On the contrary, their vision of academic language seemed to be embedded with a sociocultural understanding of learning and teaching. Clearly, the teachers understood the nuances of language and how the meaning of words is constructed in context and immersed in culture. For instance, Karmen reflected on the importance of valuing students' experiences outside of school contexts and simultaneously helping students develop concepts in their native language:

> They come with a lot of knowledge, not without it. They know a lot, but they don´t know how to show it. Only if we value it and we tell them, "But you knew it, you learned it at home. Well done." But they think they don´t know anything. And that´s not the case. On the contrary, many of these kids know how to go shopping at the Flea Market, they know how to deal with money, how much things cost. Maybe they don't know the quantities, but they know how to make tamales and sell them, and who has more or who has less. But they believe that is not education. (Karmen's Interview, translated from Spanish, Spring 2008)

★★★

> As I was telling you, in the beginning they [the students], many of them, would point to what they wanted but they would not say it. They did not have the words or they did not know the name of things, but not now, now they have to say with words what they want or what is happening to them. (Karmen's Interview, translated from Spanish, Spring 2008)

Karmen used CGI as a framework not only to understand students' thinking but also to scaffold students' thinking and push for details in their use of language. Covering teacher moves is beyond the scope of this chapter, but we have documented clearly the ways that kindergarten teachers support the development of students' thinking in other studies (see Turner et al., 2008, 2009).

Elba echoed Karmen's concern for building mathematical concepts in the native language. She states:

> Well, I really see a very strong relationship and the way I feel after doing it [CGI] this year is that the problem solving needs to be done in the students' native language then let them transfer into the second language. To me it's crucial.... I know you need to learn English and my students need to learn English, but it's going to be way faster, if I do it, if I develop that problem-solving concept in the native language and let them transfer into English. If I try to develop in English, even if I use ESL techniques and I am the best ESL teacher, just by nature it is going to take way longer because the brain doesn't absorb a lot of the, you know, the problem solving has a lot of language involved and not just social language but high cognitive language, and it takes four to seven years in quality bilingual programs for these children to develop a very good level of academic language. (Elba's Interview, Spring 2007)

In the following quote, Elba describes how she adapts this highly cognitively demanding language in problems that involve comparisons between numbers (i.e., Marios has five cars. Rebecca has nine cars. How many more does Rebecca have than Marios?).

> How many more was very difficult. I went to "how many are you missing?" to get the same as the other person. We use a lot of that, but when we rephrase it we say "how many more" until the kids realized that they were the same. And culturally, "how many more" we don't use it that much as, "I need five more to have"

like the other kids. And I think once they internalize the concept, at the end I didn't have to say, ¿*Cuántos faltan?* (How many are missing?) I just said, "How many more?" The kids had internalized it but I had to rephrase it in the beginning just for the input, the understanding input, even in their native language. Think how difficult it is even in their native language! Some words are going to be difficult so they need to find meaning for those words before they can work out those problems. Once they find meaning and they understand, they can do it. Once the procedure is understandable to them, they can generalize it and I think that's what happened. (Elba's Interview, Spring 2006)

Elba's awareness of how culture impacts meaning provides her with resources to integrate in the construction of the mathematical discourse in her classroom, helping students differentiate the nuances of language while they internalize the mathematical concept of "how many more." Ron (1999) refers to the same issue that this type of language (i.e., "how many more") poses for Latina/o students when they are exposed to compare-type problems.

Considering Culture While Adapting Curriculum

We contend that professional development needs to be grounded in teachers' practice and curriculum implementation in order to impact students' achievement. Recognizing teachers as agents of their development, their practice, and the implementation of curriculum has been discussed (Drake & Sherin, 2006). This is particularly important when trying to make sense of bilingual teachers' growth as they engage in adapting a mathematics reform curriculum to fit Latina/o students' learning needs, especially considering different aspects of language and culture that become intertwined with the teaching and learning process.

We approached our collaboration with the teachers with an emphasis on infusing a CGI perspective on problem solving into the reform curriculum as well as considering issues of language and culture in relation to Latina/o students' mathematical understanding. Our conversations with the teachers involved discussing how this particular approach to problem solving in elementary grades fit together with the curriculum.

The teachers' reflections illustrate their insights on this matter. They saw integrating contextualized problem solving into reform curriculum as a positive and necessary addition because they considered the curriculum deficient in terms of the quality and quantity of problems offered for students to solve. They commented:

> [The mathematics curriculum] does not work a lot with story problems. They give one example, which is very simple, and then the kids get hooked on that example. They all do the same example with different animals and different people but the same thing. (Norma's Interview, Spring 2007)

★★★

> And I felt that the CGI is...what my program was missing for the higher thinking skills. (Elba's Interview, Spring 2006)

The teachers' years of experience teaching a reform curriculum, together with the training in which they had participated in order to learn about its implementation, provided them with the tools to assess it in relation to what they perceive are students' learning needs. These experiences have impacted their self perception, allowing them to define themselves as authority in relation to curriculum implementation:

> When I first started, I would follow what the book says and as I was learning more, I realized what was more important. It's like, "Oh, this is really not that important, it is OK" and I will touch on it, but it's not crucial, so I made my own decisions of what I think is more important. (Norma's Interview, Fall 2006)

★★★

> I told my team, my colleagues when we were working doing the CGI that we were really lacking creative ways to problem solve, a variety of ways to problem solve, for their minds to really fly. And that became my priority this year, to problem solve. (Elba's Interview, Fall 2006)

We acknowledge that teachers' stance is not only the outcome of our collaboration, but it is an essential element in these bilingual teachers' growth path and in their decision to participate in learning about how an instructional emphasis on problem solving and communication can affect Latina/o students' mathematical understanding. These teachers understood the challenges of linguistically and culturally contextualized mathematical learning and recognized the need to scaffold children's thinking so that students value mathematics in their everyday lives. They were aware of and understood the challenges, but the professional development afforded them the opportunity to think and reflect systematically about the nuances of these challenges, connecting these reflections with theory, becoming more articulate about them, and moreover, comparing observations and discussing them with peers.

This understanding allowed teachers to see a disconnect between certain tasks proposed by the reform curriculum and Latina/o parents' background knowledge (Civil, 2002). They provided many examples that illustrate this disconnect. For instance, a homework activity asked parents and students to measure elements using cooking measurement tools such as a cup and a scale. Students returned the next day with an incomplete assignment due to the fact that it is not a common practice in Mexican families to use measurement tools. They rely on estimating the quantities required for each recipe, referred to as calculating *a ojo* (eyeballing).

This type of experience provided teachers with an important insight into the curriculum that despite its Spanish translated version still presented students and parents with linguistic and cultural barriers that they needed to mediate. In this process, we found that teachers are progressively reconceptualizing their teaching role as mediators between home culture and curriculum. Teachers' attempts at bridging the distance between family funds of knowledge and curriculum requirements translated in different ways to promote more parent involvement. For example, Norma carefully looked over the homework designed under the curriculum and adapted it when necessary to facilitate parents' understanding, making it possible for them to help their children to complete the task. She also encouraged parents to participate in early morning activities and welcomed them to observe or collaborate in different tasks. Nora decided to design and implement a mathematics workshop for parents to demonstrate how they could create problems for children to solve at home. She presented this idea to us explaining that she realized how different the actual approach to teaching mathematics is from what Mexican parents had experienced as learners. Her goal was to provide them with tools to understand the importance of mathematical reasoning over procedural learning. Norma regularly invited parents to be part of her classroom literacy activities. However, our work with her highlighted the importance of the experiences children have at home, and she was able to extend this to think about the need to provide parents with an opportunity to explore these types of problems in a workshop format. On several occasions, the invitation to parents was extended to the entire school through a parents' school organization. Norma, Karmen, Elba, and the authors were all part of this effort, with teachers taking the lead in presenting different problem types using CGI as a framework to help parents pose mathematics problems to their children and to help both the parents and their children learn mathematics with understanding (Hiebert & Carpenter, 1992).

We believe that central to bilingual teachers' development is the understanding of students' culture and the validation of families' knowledge and language. Research has demonstrated that central to enhancing students' learning is the instructional integration of cultural practices, and family and community knowledge (e.g., Civil, 2006; González et al., 2005). Norma, Karmen, and Elba understood that in order to empower their students (Cummins, 2001) to learn mathematics they also needed to include their parents, acknowledging what they know, but, at the same time, providing them with resources that supported students' academic achievement. Even though these teachers knew parental involvement was a central component of children's learning outcomes and the importance of funds of knowledge in teaching, before our professional development work, what is critical to highlight here is that teachers were working with parents in language arts, not mathematics. An outcome of this professional development experience was empowering teachers to find strategies to involve parents in a concrete manner with their children's learning experiences in mathematics. There is a difference between knowing that something is important and taking action on it.

Impact of Professional Development on Teachers and Students

Confidence in teaching mathematics. All teachers discussed the impact that professional development focused on integrating student mathematical thinking, language, and culture had on them and their own teaching of mathematics. For example, in the following quote, Karmen refers to her increased understanding of mathematics: "Before CGI, I had more questions. I had more questions about how I was teaching mathematics, but after CGI, I think I'm understanding mathematics better myself, because I never was a good student in math" (Karmen's Interview, Spring 2007).

In addition, Norma and Elba discuss how they are now affording better opportunities for students to engage in solving challenging mathematics problems:

> I would say...working the language part, you know, working a story or a problem and...having the children reason it, work it

out. I mean the whole thing, what we have been doing you know, like explaining the problem, maybe representing in a picture and with a number and the words.... I think to me that has been very valuable in my teaching experience and then I feel more comfortable doing this...than before. (Norma's Interview, Spring 2006)

★★★

I wondered if my kinders...I always believed in their ability, but I thought they are so low, I knew I could take them through addition...but division and multiplication I wasn't sure. But they were ready, they were ready. And that's what I really learned. Problem solving works, and CGI works with very young students and it works really, really well....I am very happy...because it helped them to strengthen their abilities, according to what they can do. They really have come a long way. (Elba's Interview, Spring 2006)

All teachers expressed how the professional development that focused on integrating CGI into the curriculum helped them to become more confident in their own teaching and raised their awareness of providing Latina/o students opportunities to experience higher-order thinking with challenging mathematics problems (Flores, 2007). More importantly, teachers' confidence in their mathematics teaching impacted students' learning, providing students with opportunities to develop capabilities to solve problems (Philipp, 2007; Thompson, 1992).

Conclusions and Implications

In this chapter, we focused on elementary bilingual teachers' reflections on using students' native language and culture when teaching mathematics. Scholars have emphasized the importance of building academic language in students' native language as these students concurrently transfer mathematical concepts to English. (Cummins, 1986, 2001; Thomas & Collier, 2002). We realize that having bilingual classrooms in this southwest state is a luxury that may not exist in other states. In such cases, teachers should still value the use of students' native language to communicate mathematical thinking (see Gibbons, 2009). The use of the native language allows students to internalize mathematical concepts by sharing their problem-solving strategies, thus making different strategies accessible to all students. Access to students' native language to communicate mathematical thinking and reasoning is critical and is an equity issue in mathematics education (NCTM, 2008).

Professional development with teachers should focus not only on teaching and learning mathematics, but also on integrating language and culture. We concur with Rousseau and Tate (2003) and Secada (1989, 1995) that future research and professional development models in mathematics education should place students' linguistic, racial, ethnic, and socioeconomic background at the center. Moreover, in agreement with Philipp (2007), we contend that research on bilingual teachers' professional development can contribute to recognizing teachers' knowledge and reflections and how they play a role in their everyday teaching. Our study showed that these teachers had an improved understanding of how to integrate language, culture, and mathematics. More importantly, teachers became more aware of young Latina/o students' capabilities to learn challenging mathematics problem solving and gained confidence over time in their own abilities to provide students with adequate opportunities to learn.

Note

1. The names of all participants have been changed to protect their identities.

CHAPTER 2

Centering the Teaching of Mathematics on Students: Equity Pedagogy in Action

LAURIE H. RUBEL

The preparation and training of mathematics teachers to teach in "urban" contexts is one of the most pressing issues in mathematics teacher education. The term "urban," however, takes on multiple meanings in American vernacular as well as educational research discourse. Sometimes it is used as a placeholder for race, so that students from underrepresented groups become "urban youth," regardless of whether or not they live in a city. Other times, the term "urban" is used to connote a high-poverty or high-crime area. In this chapter, the term "urban" specifically indicates "of or relating to a city." Defined in this way, urban-ness of context is significant for mathematics teaching and learning because of its inherent challenges and often overlooked benefits.

Urban schools possess two defining features, which are consequences of being located in places with high population densities, and are typically only cast in terms of the challenges they pose to teaching and learning. First, urban schools are part of very large school districts that are characterized by bureaucratic leadership structures, emphases on standardized testing, high teacher turnover rates as well as a hodgepodge of certification, induction, and mentoring programs (Weiner, 2000). Second, urban school districts serve a diversity of students, along racial, linguistic, religious, and socioeconomic dimensions, within a range of communities with varied local histories.

Of course, we can also view these essential features in terms of the benefits they might provide to teaching and learning. As cultural and economic centers, cities offer accessibility to a wide range of informal learning resources, familiarity with multiple modes of transportation, diversity of architectural spatial arrangements, and support in the form of established local communities. The linguistic, religious, socioeconomic, and racial diversity, which is often described in terms of the challenges it poses for schools and for teachers, also implies the potential for a similarly diverse teaching force as well as a diverse collection of cultural contexts from which to draw upon in the schools.

This chapter describes the initial two years of an ongoing project in the most populated and most densely populated North American city, New York City. This project, Centering the Teaching of Mathematics on Urban Youth, is a university-based high school teacher professional development and research program for urban high schools organized around the theme of culturally relevant mathematics pedagogy. I begin this chapter by broadly defining culturally relevant mathematics pedagogy and offering readers connections to related research literature. Then, I describe in more detail, a theme of maps, communities, and mathematics that extended over these two years of the project. Finally, I draw upon these results to pose directions for further research and suggest implications for mathematics teacher education and professional development.

Culturally Relevant Mathematics Pedagogy

Culturally relevant mathematics pedagogy is a pedagogical framework that builds on the ideas of Ladson-Billings (1994a, 1995) and Gutstein, Lipman, Hernandez, and de los Reyes (1997). It consists of three components: (a) advance all students toward academic excellence by teaching for understanding, (b) include aspects of the lived experiences of students and their communities as contexts for mathematization, and (c) use mathematics to describe or analyze societal inequities to develop students' awareness and mathematical literacy so that they can fully participate in a democracy.

Teaching for understanding implies an orientation toward teaching mathematics that emphasizes the connections between mathematical concepts, procedures, and facts (Hiebert & Carpenter, 1992). It necessitates meaningful classroom tasks as well as tools that support the construction of connections across concepts. Since understanding also implies mathematics as a valued activity to engage in sense making of problematic situations (Boaler & Greeno, 2000), classroom norms for participation are significant. In particular, patterns of classroom discourse, the teacher's attention to student thinking during the lesson, and the ways that the teacher makes connections to the knowledge that students bring to instruction are all benchmarks of teaching for understanding.

The second component of culturally relevant mathematics pedagogy is the inclusion of aspects of the lived experiences of students and their communities as contexts for classroom mathematization (Martin, 2000; Moses & Cobb, 2001). Specifically, this implies the use of meaningful real-world contexts as sites for mathematical representation, exploration, discovery, and communication in the classroom. As Tate (2005) explains, "One barrier to an equitable mathematics education for African American students is the failure to "center" them in the process of knowledge acquisition and to build on their cultural and community experiences" (p. 35).

The third component of culturally relevant mathematics pedagogy is to utilize mathematics to analyze or describe societal themes or inequities to develop students' awareness and mathematical literacy so that they can fully participate in a democracy. The notion of students using mathematics as an element of a voice of dissent corresponds with the literature on critical mathematics education (Skovsmose, 1994) and teaching mathematics for social justice (Gutstein, 2003, 2006). Gutstein provides evidence that teaching students to "read the world" with mathematics develops their sociopolitical consciousness and helps to change students' orientations toward mathematics.

Teacher Learning

People often use the term "culture" to refer to students' ethnic or racial backgrounds. Instead, I focus on the ways that students participate in their various communities (Cole, 1996). As Gutiérrez and Rogoff (2003) explain, this stance represents a "shift from the assumption that regularities in groups are carried by the traits of a collection of individuals to a focus on people's history of engagement in practices of cultural communities" (p.21).

The implication of "culture" and "relevant" in culturally relevant mathematics pedagogy is that teachers need to have knowledge about their students and their students' communities. They need to have a sense of the local community, its history, the activities their students engage in outside of the mathematics classroom, and how they participate in those activities. This knowledge about students is necessary to be able to (a) create classroom norms that support student participation necessary for developing mathematical understanding, (b) select meaningful and relevant contexts for mathematization, and (c) identify (or guide students toward identifying) social issues that can be analyzed or described with the mathematics at hand. The professional development efforts described in this chapter were oriented around building these aspects of teacher knowledge while also expanding upon the teachers' mathematical and pedagogical content knowledge.

Project Structure

The professional development was organized primarily around teacher learning as a group—participating teachers attended a week-long summer institute, and then continued to meet, as a group, in seven monthly meetings across each intervening school year. I also visited the schools and classrooms of participating teachers and interviewed each individual teacher at least four times as part of the project. This chapter describes two iterations of this cycle: the summer institute of 2005, the 2005–2006 school year, the summer institute of 2006, and the 2006–2007 school year.

There were seven participating teachers in the 2005–2006 cycle, followed by 11 teachers in the 2006–2007 cycle. All of the teachers individually elected to participate in the project: five teachers participated in both cycles, and it is two of these five teachers that are profiled in this chapter, Mr. Red and Ms. Purple[1]. Both Mr. Red and Ms. Purple had five or fewer years of teaching experience at that time and had become mathematics teachers through an alternative certification program; in that program, they had studied in my education courses about pedagogy and action research. Both teachers were in their 20s, had accumulated enough undergraduate mathematics credits to qualify for and eventually earned state Grades 7–12 teaching certification. Mr. Red is Asian and Ms. Purple is White; both were relatively new to New York City.

Mr. Red taught at a high school for recent immigrants to the United States, of about 475 students, in which about 75 percent of the students were eligible for free lunch and about 60 percent were classified by the school system as "Limited English Proficient." Close to 60 percent of the students self-identified as Hispanic or Latino, and another 30 percent as Asian. I visited Mr. Red's school six times during

2005–2006 and four times during 2006–2007. I also conducted a total of eleven 30–50 minute interviews with him.

Ms. Purple taught at a high school that was opened in 2004 on the campus of a large high school that had been closed due to poor performance. With about 350 students, about 75 percent of the students were eligible for free lunch, and about 9 percent were classified by the school system as "Limited English Proficient." Close to 60 percent of the students self-identified as Hispanic or Latino, and another 25 percent as Black or African American. I visited Ms. Purple's school four times during 2005–2006 and once during 2006–2007. I also conducted a total of four hour-long interviews with her.

Both Mr. Red's and Ms. Purple's schools primarily serve students from low-income families, and both schools have high proportions of Black or Latino students, a combination that is typical of urban schools and of the schools involved in this project. It is important to remember that the project's urban context implies tremendous diversity across the participants' schools. While these particular two schools are both majority "Hispanic or Latino" schools, this signifier does not necessarily imply homogeneity within or across the schools. This broad category includes, for instance, third-generation New Yorkers of Puerto Rican background and first-generation immigrants from Ecuador. Similarly, the category "Black or African American" signifies a similar heterogeneity in this local context in that this category encompasses Afro-Caribbean, dark-skinned Latin American, and African American students.

Summer Institutes

The primary goal of the 2005 summer institute was to develop a professional community of learners, with a focus on equity pedagogy in secondary mathematics. We read selections from professional literature (Ladson-Billings, 1994a; Moll, Amanti, Neff, & Gonzales, 1992; Tate, 2005) and looked for ways to relate the various notions of culturally relevant pedagogy, the process of drawing on students' funds of knowledge, and pedagogy that centers on the lives and experiences of students. We also investigated examples of connecting mathematics to societal themes or inequities. For instance, we examined an application of the area under the curve to find the Gini coefficient (see Staples, 2005), a measure of economic equity. A second activity involved the plotting of data of U.S. minimum wages over time, as a context in which to discuss rates of change. We modeled that data with appropriate mathematical models. We also refigured these wages in current dollars using the Consumer Price Index, so as to reflect inflation, all as a way to consider and analyze the "Living Wage" movement.

In the 2006 summer institute, five of the seven participants from the previous year continued to participate, and were joined by six additional teachers. The goal of this institute was for teachers to explore using mathematics with students to critically examine their world. Participants read selections from Gutstein (2006), and Gutstein cofacilitated one of the sessions. Sessions were designed with the goal of providing teachers with experiences and potential tools with which they could identify mathematical themes. Some sessions were led by returning teachers, "old timers" (see Stein, Silver & Smith, 1998) in the community, in which they presented detailed examples from their own instructional practices in the form of lesson plans, student work, and reflective comments.

Maps, Communities, and Mathematics

Maps as Representations

One theme that extended over the span of these two years of professional development was the use of maps as mathematical representations of geographical communities. Maps can be connected back to the framework of culturally relevant mathematics pedagogy in the following ways. To start, maps as representation raise the notion of the importance of representation of mathematical ideas, and the usefulness of these ideas as a tool in building understanding. Maps of local regions are also relevant for students and can be used as contexts for various types of problem posing. And finally, maps of local communities can function as tools with which to explore, identify, represent, or challenge statements about local inequities.

In the professional development project, we began with iconic maps from our own city and examined them critically, practicing a question of identifying the point of view of the representation—how is the map scaled and according to what variables? What story does this map tell about this place, and what other stories might one be able to tell? Together, we created cartograms, which are maps based on scales other than land area. For instance, when we represent the space using a scale like average family income, different land areas of the city become larger than others. Our cartograms used such variables of scale as human population, dog population, number of public recreation spaces, or number of Starbucks branches, generating different representations of the same land area. One can look for relationships across different variables of scales, relative to a certain neighborhood, or one can use these relationships to contrast different neighborhoods.

Participating teachers were quick to notice a variety of ways to connect local maps to mathematical concepts that are part of the high school curriculum. Descriptive statistics can be calculated using data easily accessible from U.S. Census or other state or local government agencies. Maps, as scalar representations, can be used to explore concepts of ratio, proportion, and area. Maps can also be used as contexts for geometric notions of center (means or medians), such as finding the population center of a specific geographical area and comparing that to that area's land center. Alternatively, maps can also be used to create vertex-edge graphs, as a way to explore numbers of possible routes, circuits, or minimal paths. Such graphs are useful tools in analyzing issues of fairness related to public or competitive business facility location.

In addition to examining local maps of our own city, we also considered the mathematical problem of how to map the Earth, a sphere, onto a two-dimensional paper. Drawing on Gutstein (2006) and Wood, Kaiser and Abrams (2001), we considered the choices inherent in creating such a two-dimensional representation: Do we prioritize preserving angle relationships between places, crucial for maritime navigation, or do we prioritize other relationships? For instance, the commonly used two-dimensional representation of the Earth, the Mercator projection, preserves angle relationships between locations. But in doing so, it distorts other features of the space, like relative area. Places further away from the Equator appear relatively bigger using the Mercator projection than they do on the globe itself. A less common representation, the Peters projection, scales the land, instead, according to area but distorts angle relationships between locations. Some parts of the world look similar on these two maps; other parts of the world appear very different in relative size. Teachers quickly noticed that, since each map describes different features of our world, it is crucial that people be able to read maps with a critical perspective of purpose.

Maps and Communities

A second, and perhaps competing, way we worked with maps was using the tool of community walks. Since all of the participating teachers, including the teachers of color, were outsiders to the communities in which they taught, this implies the necessity to learn about their students, their students' families, and their students' communities. Thinking back to the framework of culturally relevant mathematics pedagogy, teachers need to build a working knowledge base about their students, their various activities outside of the mathematics classroom, and the ways that their students engage in those activities. I created a community walk activity for the teacher professional development with this goal in mind.

In the first iteration of the community walk, in the summer of 2005, the teachers were grouped in pairs, and each pair received a street map of a distinct Census tract, all in the immediate vicinity of the university campus, in a low-income, Afro-Caribbean neighborhood. The teachers all had extensive experience traveling through this neighborhood, but none of them lived or taught there. Teachers were given the option of (a) choosing a particular theme a priori (such as access to financial institutions, diversity of grocery or food options, availability of recreation spaces, or indicators of gentrification) and then walking through the tract to investigate that theme, or (b) walking the tract with an open mind to allow potential issues in that neighborhood to emerge. That afternoon, teachers recongregated as a whole group to share their findings. They were then introduced to a variety of electronic resources that contain information that is directly linked to those Census tracts, such as the Census information, or city and state data about housing, public health, education, and other themes. They also investigated various websites (c.f., www.socialexplorer.com) that display time series maps linked to categories contained in U.S. Census dataset.

In the 2005–2006 school year, the teachers built on their experiences in the professional development, and with the mapping activities in particular, in different ways. For example, Mr. Red created a project titled the Community Mapping Project, in which his students conducted their own individual community walks and gathered data about their own neighborhoods according to variables of scale of their own interest. They constructed scale maps of their own communities and retrieved and analyzed local Census data. Each student also compared some aspect of his or her neighborhood with that of a classmate and made predictions about the future of their neighborhoods. Two students mapped businesses and locations in their neighborhoods and scaled each map according to how important each location is to local residents.

In Ms. Purple's case, she designed a variety of projects around the context of the local city. In one project that semester, students worked on graphing, modeling, and predicting local populations, using algebraic models. Their initial data for this project was drawn from the U.S. Census about their own Census tract. Another project had students devise the number of digits needed for license plates in fictitious municipalities on the basis of their home Census tracts. When asked about student engagement in these projects, Ms. Purple explained,

> They liked working on stuff that's about them.... I don't know if they liked working on stuff about New York City per se, but they definitely liked working on stuff which was all about them and what they would do and how would you reflect yourself in something. (Interview, 4/06)

While I intended the activity of a community walk to be a tool with which teachers could learn about their students, by the end of the 2005–2006 year, I noted how this tool was being used differently by the teachers. They seemed to have primarily gleaned from this activity that they could use maps, or physical aspects of neighborhoods, as contexts in which to explore various mathematical concepts, one of the three components of culturally relevant mathematics pedagogy. Although these and other curricular projects successfully implicated the students and their communities by using local contexts, which was certainly a goal of the Centering the Teaching of Mathematics on Urban Youth project, the teachers seemed to be solely focused on changing the mathematical tasks in their lessons and not on also getting to know their students and how their students engage in activities outside of the mathematics classroom.

In a second iteration of the community walk, in 2006, the following summer, the teachers were still paired; each pair chose one of the same census tracts, again in a low-income, Afro-Caribbean neighborhood, but each pair was asked to adopt one of the following data collection strategies:

1. Find a spot that is interesting to you. Sit silently and observe for a period of about half an hour, without interacting with anyone and without taking notes. After this time of silent observation, make detailed notes of what you observed. Then discuss your findings with your partner.
2. Choose a theme with your partner. Walk the tract and make notes about your theme as it relates to this neighborhood. Discuss your

observations with your partner as you go along. You may want to create some sort of graphical representation of your finding.
3. Choose a theme with your partner. Try to talk to as many people as you can to gain information about the theme as it relates to this neighborhood.
4. Walk your tract. Take notes of things of interest as you go along. Then create a graphical representation(s) of the tract that reflects what you find interesting about the neighborhood.

This time, I participated in this iteration of the walk; paired with another facilitator, we identified establishments within this Census tract that seemed to be typical of low-income, urban neighborhoods, about which we knew very little: a pawn shop, a check cashing store, and a rent-to-own furniture outlet. Neither of us had ever visited these types of businesses before, so this represented an opportunity to try to understand how these stores operate by talking to people engaged therein. We wanted to find out who seems to frequent these particular stores in this neighborhood and to gain an understanding of the local role of these businesses. We shared these experiences with the teachers in the debriefing session after the walk activity. Curiously, again, the teachers, on their own walks through the neighborhood, did not interact with anyone they encountered; during either summer, only one teacher, one of the African American participants, reported talking with anyone on the three-hour walk.

In the 2006–2007 school year, Mr. Red revised his previous community-mapping project into a new project, which he titled the Atlas of Origins. His students, all recent immigrants to the United States, created scaled maps containing the countries of origin represented by all of the students in the class. Each student's map was to be drawn with the map literally centered on that student's own country of origin. Ms. Purple also continued with the theme of maps and had students create scale representations of the city's five boroughs, scaled according to a variable of their choice, representing a public health or social issue.

Reflections

I intended the professional development activities related to mapping to serve two purposes. One, I hoped that working with maps as a way to represent ideas about communities would function as an example of how to ground mathematical explorations in students' local, lived experiences. Ms. Purple and Mr. Red, as examples of the group of the teachers, created and implemented projects that related to maps, as described earlier.

The second purpose of the work with maps of communities in the professional development was for teachers to use the activity of a community walk as a tool for their own learning about their students. This learning could be useful in terms of building relationships with students, improving classroom participation structures, or identifying other relevant contexts that could be explored with mathematics. Even though we practiced this activity together as a group over two consecutive summers, teachers actively took up mapping as a context for curriculum but resisted the ongoing suggestions, in the summer institutes and in the professional development meetings, to use community walks, or other tools, to develop their own knowledge about their students. Although we explicitly discussed shadowing a student at school, making a home visit as described by Moll and his colleagues (1992), visiting a neighborhood church, or having a student lead a teacher on a neighborhood "tour," these tools were not taken up by the participating teachers in these two years of the project. Readers interested in one teacher's perspective about this professional development project can refer to Chu & Rubel (2009).

One possible interpretation of this difficulty for teachers is that they viewed the purpose of this professional development as a way to address their immediate teaching needs, to generate ideas for curriculum. Participants were primarily beginning teachers, so their focus on curriculum is entirely reasonable and sensible. A second way to view this difficulty is in terms of the general tendency of high school teachers to view teaching as conveying their own relationship with their subject area to others (McLaughlin & Talbert, 2001). In this case, the participating teachers' initial interpretation of culturally relevant mathematics pedagogy was to use mathematics as the starting point and find creative ways to connect that mathematics to their students' physical worlds. My goals in working with them to develop culturally relevant mathematics pedagogy was challenging them to change their conception of teaching to be, instead, the building of relationships with students, and among students, about mathematics (Lampert, 2003).

Race and Class as Salient Issues

Another possible interpretation of this difficulty for teachers is that as outsiders, they would have to negotiate difficult "border crossings"

(Anzaldua, 1987) to gain a deeper knowledge of their students. Teachers cross borders on multiple levels. First, teachers belong to a different generation than their students and know more mathematics than their students. Many also come from a different geographical region than their students, live in different parts of New York City, and in most cases, speak different home languages or dialects than their students.

Perhaps one of the most significant "borders" between teachers and students was their different racial-ethnic and social class reality, complicated, of course, by urban diversity. One of the participating teachers, an African American, explained,

> I've always thought as far as racism or prejudice is concerned, it was Whites against everyone else. But some of the things that are wedging between Blacks and Dominicans, and Dominicans and Puerto Ricans, and Puerto Ricans and Blacks, and Blacks and— I'm amazed.(Interview, 9/06)

These different racial realities clearly play an important role in shaping how mathematics might be relevant, both for whom and for what purposes. For instance, Ms. Purple explained,

> Sometimes I feel like there's a disconnect between what I think is important and what they think is important. And when I'm looking for things, I've got my view of the world looking at things and I may not be looking at things the same way. But also, sometimes, I feel like there's this inherent distrust on the part of the students for somebody of a different race. And so I'm somewhat limited to what I can do because I might reach their limit. (Summer Session, 8/05)

Oftentimes, issues of race, ethnicity, or social class are pointed to or discussed among teachers in professional development sessions, but integrating aspects of students' lives into mathematics instruction invites these issues into the classroom in explicit ways. For instance, the school projects related to mapping brought discussions related to race and ethnicity to the surface. In some cases, the general racial categories from the U.S. Census did not correspond with students' categories of self-identification. Ms. Purple described,

> And then we had a discussion, well, what is White? And they're like, "Well, you're White." And I said, "Well, what makes me

White and not you White?... We both have olive skin tones and we sort of look alike, we have the same color hair, but I'm White and they're not."... And even my Black students, they didn't like it that it was Black or African American because, you know, this one girl goes, "But I'm from Jamaica." And I said, "Well, that puts you in the Black category." She goes, "No, I'm Jamaican. That's what I want to be." (Interview, 4/07)

On one of my visits to Ms. Purple's classroom, an Ecuadorian American student, looking at the categories on the computer screen, asked me if I knew what "Hispanic" meant.

Yet these discussions seemed to distract from the mathematics objectives of Ms. Purple's project, instead of being front and center. In other words, Ms. Purple's lesson did not include the posing of mathematical questions about this very issue of the Census racial categories. For instance, how have the categories changed over time? How might a different distribution of racial categories lead to a different description of a given location? The students, on their own, noticed the category of "Other" and wondered about who might identify that way. Again, mathematics could be used to analyze how the category of "Other" has changed over time, how the changes in the frequency of the "Other" category reflect other changes in the city. This analysis could inform an examination of the possible implications for communities with large numbers of people identifying as "Other."

Students' negotiations of and about the U.S. Census racial/ethnic categories brought matters of race (and the racial borders between students and teacher) to the surface in Ms. Purple's classroom. However, these racial and social class differences played a role in the mathematics lessons in more subtle ways as well. For instance, in one of the lessons designed by Ms. Purple, students needed to access and retrieve population data from their home Census tract. This is easily accessible to individuals on the Internet, provided that one enters an address in the form recognized by the Census. In this particular urban context, people typically use an alternate form of identifying locations, by naming intersections, and not using street numbers and zip codes. In planning her lessons, Ms. Purple assumed that her students' knowledge of their home locations would conform to the Census format and therefore did not anticipate their ensuing difficulties with her planned lesson activities.

Final Thoughts

During February of the 2006–2007 school year, at one of the group's monthly meetings, I raised the distinction between creating curricular projects for students involving maps of communities and teachers using community walks or other tools to learn more about their students. I asked the teachers why they were only engaged with the former and seemed resistant to engage with the latter. Some of the teachers responded immediately, and perhaps defensively, that they did not see how this could relate to their teaching of mathematics. Others indicated that they sufficiently observe the students in and around school. Still others quietly absorbed this feedback.

A month later, in a follow-up group meeting, Mr. Red commented on a newly discovered distinction,

> I still think in terms of knowing things about the world that might develop in the students rather than knowing things about the student...Even "centeredness," I've taken in my work as "let's make it about you" rather than "let's find out about you."...I mean, that's the question, right, where am I getting my knowledge? I don't know. It is in classroom settings, it's by sources other than going to their communities. (Group Meeting, 3/07)

> Mr. Red continued to reflect, "It's the difference between seeing and doing. I mean, you're asking us to really—we're comfortable with doing things, writing lessons, teaching them. But to really be, in a different way, is hard" (Group Meeting, 3/07).

This experience has taught me that teaching teachers about culturally relevant mathematics pedagogy consists of two intertwined strands. Teachers need to learn how to "find" mathematics in everyday situations and how to pose mathematical questions about those situations. The focus on maps described in this chapter demonstrated a way for teachers to engage in learning about that process. In the current environment of high-stakes testing and district pacing calendars, the challenges inherent in thinking about curriculum development in this way should not be underestimated.

The second, inextricable strand involves knowing one's students, being able to select contexts that are relevant to their lives, and organizing instruction so to maximize students' participation in the lesson. In other words, working with teachers on culturally relevant mathematics pedagogy cannot just focus on how to create curriculum that contextualizes school mathematics in experiences that are relevant to one's students. Alongside this endeavor must be an effort to teach teachers about the importance of building relationships with their students and learning about their students as a necessary part of this process. As Mr. Red explained, "I failed to realize that being culturally relevant was not possession of some body of knowledge, but rather is a matter of relating to students, being involved with them, and engaging them" (E-mail, 5/07).

The two years of professional development described in this chapter focused only on working with and supporting teachers and not on researching the effectiveness of this approach for these teachers' students. In an ongoing iteration of the Centering the Teaching of Mathematics on Urban Youth project, with a new cohort of participating teachers, I am now looking to closely examine learning outcomes of culturally relevant mathematics for their students. In particular, in addition to providing professional development and studying teacher learning, my research is now also examining the ways that participating teachers and schools can create opportunities for students to participate in mathematics and the ways that students take up these opportunities.

Note

1. The names of the participants have been changed to protect their identities.

CHAPTER 3

The Power of One: Teachers Examine Their Mathematics Teaching Practice by Studying a Single Child

MARY Q. FOOTE

The growing diversity of students in U.S. classrooms is met with a teaching force that continues to be over 80 percent White and middle class (Howard, 1999; Nieto, 2004). Research indicates that White teachers often have difficulty relating to children who are not White and middle class. A close examination of a child from a nondominant group may provide a base from which White middle-class teachers can develop a sensitivity that can support them in better understanding students who come from backgrounds different from their own.

Much has been written describing professional development efforts that use student work and incorporate a focus on student thinking in mathematics to support teacher growth and development as teachers of mathematics (Ball, Lubienski, & Mewborn, 2001; Carpenter, Fennema, & Franke, 1996; National Research Council, 2001). Because children's mathematical thinking develops in the multiple contexts of their lived experience both in and out of school, careful consideration of children's mathematical thinking and the multiple social contexts in which it develops may support teachers in becoming better teachers of mathematics for all children, including those students from nondominant groups. The goal of this study was to investigate how participation in a study group, in which teachers each explored the mathematical thinking and in- and out-of-school experiences of an individual child, contributed to teachers' growth and development as teachers of mathematics.

Enlarging the Knowledge Base for Teaching Mathematics

Ladson-Billings (1994a; 1994b; 1995) proposes that it is necessary to understand and to build on the community practices that children bring with them into the classroom. She argues that in addition to being mathematically substantive, instruction must also be culturally relevant. Echoing this notion, Villegas (1993 as quoted in Zeichner & Hoeft, 1996) contends that "making home visits, conferring with community members, talking with parents, consulting with minority teachers, and observing children in and out of school" are ways in which teachers can begin to understand the cultures of the children in the classrooms who are not like themselves (p. 538). These methods have supported teachers in making deeper connections with students' lived experiences.

In reviewing the work of a number of researchers, Banks (2004) concludes that "thick descriptions of the learning and cultural characteristics of students of color are needed to guide educational practice" (p. 20). In other words, it may be profitable to consider whether a close and deep investigation of children in the multiple contexts of classroom, school, and community can provide teachers with a repertoire of knowledge to draw on to better support children's learning.

Moll and associates (González, Andrade, Civil, & Moll, 2001), through their research program known as the "funds-of-knowledge" project, propose that understanding the context in which a child operates outside of school and the expertise and competencies that she or he shows in that context can be a powerful insight for a teacher that can support him or her in establishing a connection to that child. Attending to the specific educational needs of that child has the potential to support the teacher's growth and development. From the onset, the funds-of-knowledge project had as one of its explicit goals to reject a deficit theory model for the education of minority students (Moll, 1992). It adopted as its premise that it is possible to capitalize on students' and their families' experiences. In other words, the project was based on a belief in the student as a person with a broad base of valuable experiences and resources.

As teachers went as learners into these students' homes, the usual dynamic between parent and teacher was changed and the knowledge that the family possessed was honored (Civil, 1995a). Instead of the deficiencies that are often expected, teachers were supported in seeing the positive aspects of the families; parents were now seen as resources for their children. Civil points out the importance of specific knowledge about individual children in order to address the needs of each student. She notes, "A key aspect of our work is to get to know as much as we can about each individual student" (1995b, p. 13).

Professional Development that Examines the Case of an Individual Child

One professional development approach that has supported teachers in learning about particular children so as to become better teachers of those children is called the Descriptive Review Process (DRP), and involves taking an in-depth look at an individual child within the context of an ongoing study group of teachers (Himley & Carini, 2000). By embedding professional development using children's work in mathematics within the DRP, the focus of the teachers could be directed to include a consideration of other aspects of the child, in addition to her or his mathematical thinking, which could be brought into play to support her or his mathematical development. The teacher could reach beyond the mathematics classroom into the greater school context and beyond, into the home and community in order to access interests and competencies that the child possesses and which might be used in the service of her or his learning of mathematics.

Research Questions

The literature suggests that enlarging the knowledge base of teachers to include (in addition to knowledge of children's mathematical thinking) information about the lived experiences of children unlike themselves may support White middle-class teachers in their journey to becoming effective mathematics teachers of diverse populations of students. I thus arrived at a broad question that interested me: how can White teachers be supported in becoming multicultural or culturally relevant teachers of mathematics? In this study, particular attention was focused on (a) how engagement in the study group contributed to understanding a particular child as a learner of mathematics, (b) how this engagement contributed to understanding other children, and

(c) how this engagement informed the teachers' perspectives on mathematics teaching practice.

Methods

Six teachers engaged in a professional study group in which they each conducted a case study of a particular child from their respective classrooms who struggled in mathematics. To minimize issues of essentializing based on comparisons across cultural groups of students, the teachers were asked to choose an African American learner. Of course, not all children from nondominant groups struggle with mathematics. And so, why then, particularly study children who struggle? The attention in this case is on the teacher who is searching for ways to improve her practice in order to be a more successful teacher of those children who in the past she has not reached, those who continue to struggle.

The study group met for 12 two-hour sessions over the course of one semester. Ten of the 12 study group sessions were devoted to a descriptive review (Himley & Carini, 2000) of one of the target children. These ten sessions were divided into two rounds of five sessions, so that each teacher made two presentations of her student.

Participants

The participants in the study were six White in-service teachers who at that time worked in the same elementary school in a moderately sized Midwestern school district. The experience of the teachers ranged from 8 to 17 years of teaching. They taught various grades from kindergarten through fourth grade.

Preparation for the Study Group

Parents of the target children were enlisted to participate in the study. They were asked to meet with their child's teacher to discuss their child's interests, experiences, and expertise outside of school. The parents (all of whom were the mothers) were asked to bring photographs to this meeting to use as artifacts to support the discussion. Before the study group began, I met with the mothers of each of the target children to give them cameras and film and explained to them the nature of the photographs they should take of their child at home and in community settings. I gave the parents parameters adapted from the Family

Photography Project (Spielman, 2001). I asked the parents to document times when their child was (a) engaged in an activity that was particularly interesting to them, (b) engaged in an activity at which they were particularly competent, (c) engaged in a household routine such as cooking or grocery shopping, and (d) engaged in an activity that involved mathematics or attention to number.

I used photography as an alternative to the home visits that other researchers and teachers have used to enter the homes and lives of children and families (also see Allen et al., 2002). Along with being less time consuming, in this way the locus of power was shared with the family as they became the active agents in documenting the funds-of-knowledge in their homes and communities.

Study Group Assignments

Between the two presentations, each teacher met with the mother of her target student. The meetings were held either in the child's home or in the child's classroom at the election of the mother. The purpose of the meeting was for the parent to share with the teacher information about the child's out-of-school interests and experiences, using the photographs she had taken as an artifact to support the discussion. The intention was to position the parent as the one from whom the teacher would learn about the child, reversing the traditional power dynamic between teacher and parent. The meetings lasted from a half hour to an hour.

Between the two rounds of presentations, each teacher also spent one day shadowing her target child in the school setting. This provided the opportunity for the teacher to observe the child in situations such as lunch and recess, pull-out programs, and special classes such as art and gym, as well as to observe the child in the classroom itself without being a participant in the activity of the classroom. Each teacher was also encouraged to have informal discussions with her target student in order to learn from them about activities outside of school that they participated in and enjoyed. They were also asked to notice, within the school setting, the activities that the student chose to participate in when given a choice.

Study Group Sessions

When presenting a student to the study group for review, the teacher brought samples of that child's work in mathematics, anecdotal information about the child that focused on that child's mathematical thinking and problem solving, and information about that child's in-school and out-of-school interests and competencies. The teachers in this study already used in their classrooms a mathematics problem-solving framework adapted from *Cognitively Guided Instruction* (Carpenter, Fennema, Franke, Levi, & Empson, 1999). This same framework was used in the study group to analyze the strategies students used to solve these contextualized problems. In the end, other mathematics content such as counting strategies and geometry activities were also brought to the group as samples of work.

The presentation format used in the study group session was adapted from the DRP (Himley & Carini, 2000). I met with the presenting teacher before the study group session to prepare the portrayal of the child. Together we decided on a focusing question to guide the reactions of the group to the presentation. After the teacher's presentation, I opened the session to questions and comments. Each participant was asked to contribute thoughts or questions. During this time, the presenting teacher listened to the questions and comments but did not respond. Next, recommendations were presented for the teacher's interaction with and teaching of this child. The recommendations focused on the issues of the child's mathematical thinking and learning as well as on other issues that the teacher had raised in the presentation. The sessions closed with the presenting teacher commenting on what she had found particularly useful about the comments and recommendations.

Data Generation Methods

The data in the study were generated before, during, and after the study group sessions. All the pre- and postinterview as well as the study group sessions were audiotaped and the audiotapes transcribed to facilitate data analysis. Throughout the process, I took observational field notes.

Initial interview. The initial interviews with the teachers probed beliefs about what it means (a) to be successful or to struggle in mathematics, and (b) why it might be that a child would struggle in mathematics.

Study group sessions. At the end of each session of the study group, teachers wrote reflections about their evolving conceptions of the child they were studying, as a learner in general and a learner of mathematics in particular. In addition, they reflected on their evolving sense of themselves as teachers of that particular child and by extension of other children. The meetings that I held with the teachers to assist them in preparing the descriptive reviews of their students for presentation to the study group were audiotaped and transcribed.

Final interviews and writing. I conducted individual interviews with each of the teachers after the conclusion of the study group. Along with some general questions about what the participants had learned, they were further probed regarding the impact of the study group, and the interactions with the student and the student's parents, on their (a) understanding of the target student, (b) understanding of other students, and (c) understanding their own practice. At the end of the final interview, teachers were asked to write a brief description of their target child that they might give to the next year's teacher.

Artifacts that were analyzed thus included (a) transcriptions of discussions and interviews, (b) participants' writings, and (c) field notes and reflections. The use of these multiple resources means that the teachers' ideas were accessed through a triangulation of perspectives: (a) oral interviews and group discussion, (b) formal presentation, and (c) personal written reflection.

Data Analysis Methods

Using the research questions as a guide, I broadly categorized the data as to whether they pertained to issues of the particular target students, issues of other students, or issues of teaching practice. I also was sensitive to themes that emerged directly from the data, such as the participants' views of the home environment of the target children.

I applied a modified Grounded Theory using both themes that I predicted on the basis of research questions and themes that emerged directly from the data (Strauss & Corbin, 1998). To ground any shifts that occurred during the study, I examined the data with a broad lens, looking at teachers' initial points of view as well as examining shifts in the conversations and writing that were a part of the study group experience itself. I concluded this broad view by synthesizing the interviews and written responses that were generated at the conclusion of the study. Although the use of this lens provided a general sense of some of the changes that happened during the course of the study, some important particularities were not evident in this type of analysis. For this reason, I then used a narrower lens to examine the cases of two individual teachers. This perspective provided a view of the degree to which these teachers were able, not only to learn significant new information about their target children that could support the teaching of mathematics to that child, but also the degree to which those teachers recognized this learning and were able to act upon it.

Results

In the results I begin by looking at teachers' initial stances about what it means to struggle with mathematics. I then examine shifts in the conversations and writing that happened during the study group sessions themselves. I next present a specific look at the cases of two teachers, focusing on some of what they learned about their target children and the extent to which they acted on that knowledge. I then examine the interviews and writing done at the conclusion of the study.

Nature of Teachers' Stance at the Onset of the Study

There were two areas that the six teachers discussed in their prestudy interviews that were particularly pertinent to this study: (a) what it means for a child to succeed with or to struggle with mathematics, and (b) where the responsibility for school success or failure is located.

To succeed or struggle with mathematics. In discussing the nature of success and failure in mathematics, all six teachers took the question to be one of success and failure in *school* mathematics. None of them discussed the ways in which children might use or encounter mathematics in their daily lives. Issues of success were linked to making progress, or showing growth in understanding in the school setting on the one hand, and meeting standards or goals, or being successful on school mathematics assessments on the other hand. For example, one teacher (Nan) said, "I think to be successful means to solve problems, to understand what's being asked and to be able to solve them accurately and efficiently. To understand the numbers." Another teacher (Rebecca) said, "I think hitting the standards would be a way of measuring success."[1]

In a reciprocal way, struggling with mathematics was most often seen to be an issue of students' lack of ability to deal with school arithmetic or problem solving, because either there were concepts they did not understand or they lacked the necessary background knowledge. For example, one teacher (Meg) said, "[Struggling in math means] they can't verbalize to the teacher that they don't understand." Another teacher (Rebecca) in focusing on the interaction between teacher and student said, "You [i.e., the students] don't understand what the teacher's talking about. How she's trying to help you understand these groups of numbers really doesn't make sense."

Locus of responsibility. Two teachers (Caitlin and Meg) indicated that the locus of responsibility for a child struggling with mathematics might rest with the teacher or the school mathematics work, saying that the work might be too hard for the child, or the teacher had not found a way to teach that child. Although these two teachers listed teacher or school responsibility as one factor that might contribute to a child struggling with mathematics, Caitlin also indicated that home or environmental factors such as home experiences, or home language that intersected poorly with school expectations, might be contributors. All but one teacher (Meg) placed some responsibility for struggling with mathematics outside of the school and in the home of the student. The teachers began the study then, with a view of the home environment of the children they were studying, not as a resource but as a potential impediment to learning.

The Study Group: Experiences and changes. Throughout the presentations and responses in the study group, there was a tension between the discussion of the target student's mathematical learning and the discussion of the more general knowledge of the target student's interests, expertise, and experiences. At times a more general look at the child took prominence over looking at his or her mathematical thinking. Even when the teachers observed mathematics-related activities, however, several did not follow up on them or discuss them in depth.

Individual Narratives

To examine particular cases of teachers' trajectories as they participated in the study group, I present a set of two narratives of individual teachers. In doing this I focus on the varying degrees to which these two kindergarten teachers learned from the experiences of the study group and applied that learning through making or proposing changes in their classroom practice. In the case of Ellie we see a teacher who learned about her target child's mathematical performance outside of school and who adapted her practice to accommodate this learning. We see in detail how she was able to access information about the child through meeting with the parent. In the case of Tina we see a teacher who learned about her target child's interests outside of school and who recognized the potential of some of that learning to impact on mathematics instruction, but who did not act on it. Ellie stands alone among the group in the extent to which she was able to change her beliefs and practice in response to her interactions around her target student. Tina represents the experiences of the majority of the study group participants, who to varying degrees were able to recognize, although not take up, information they had gained about their target student in and out of school.

The case of Ellie. Ellie was the one teacher who across the multiple experiences of shadowing and conferencing was able not only to learn surprising things about her target student, Evan, but was also able to incorporate that learning into the classroom context in ways that supported Evan, a kindergartner. One significant event for Ellie was her meeting with Evan's mother to review the photos the mother had taken. In the meeting to prepare for her second presentation of Evan to the study group, she discussed one photo in particular that had informed her understanding of Evan's performance in mathematics:

> The picture that impressed me most from a math point of view was the picture of him counting the jewels on a transparent plastic belt that he was looking at.... I bet [the belt's] about 40" long and I bet there are five jewels to an inch practically.... His mom mentioned that he got to 30. That surprised me a little.

Ellie analyzed the interaction around this particular photo in several ways on different occasions. Initially, she wondered if the mother was overstating the child's performance since Ellie had not seen evidence of Evan's counting ability in the classroom. Later, she thought that the child's success in the context of counting the jewels on the belt might indeed exceed the performance he had shown in the school setting. She interrogated the reasons why that might be so and implemented changes in the classroom because of it. She wondered if Evan's success in the belt context might be due to the stationary nature of the objects. She speculated that perhaps the organizational challenge of using counters that move around the table and fall onto the floor inhibited the child's counting performance. With this in mind, Ellie gave Evan a counting frame to use to practice his counting skills. Although the beads in the frame still move, they are captured in groups of ten on a metal rod. They cannot fall on the floor, and double counting is more easily avoided. This is in fact what happened; Evan was able to be more successful with counting using this material. I am not suggesting that Ellie was restricting Evan's access to tools and materials in the classroom, but rather that she was examining the extent to which her provisioning of the environment supported or inhibited performance at a particular moment.

Ellie recognized that there were certain classroom structures that she had in place that supported Evan's participation in her classroom. In her second presentation she said,

> He is more engaged in the workshop format than he is in centers. And that's because he has a choice. It's a self-regulated thing, rather than having a prescribed series of activities. It takes a lot of effort for him to engage in activities that aren't of his choosing or that are new to him.

During her second presentation to the study group, Ellie spoke of changes she had made in her classroom structure in order to build on Evan's interests while supporting him in having a place in the classroom to verbalize his understandings. She said,

> I've been thinking about his love for play and how I'm not getting a lot of verbal feedback and so I've been trying to do some more structured play in the classroom. I've been trying to create situations where there are certain kinds of things to use to see if we can build some language and context of different activities to support him starting to talk a little bit more about things... so that I have a better sense of what kind of learning is taking place.

The experience of discussing with his mother the photograph of Evan counting the jewels on her belt provided access to knowing Evan in a new way. It provided support for Ellie to change her stance toward Evan. At the beginning of the study, she had viewed Evan as an inattentive, unfocused child with few academic or social skills. The problem was located within Evan. By the end of the study, due in part to the interaction with Evan's mother as well as her own observations of Evan in the school setting, Ellie saw much to build on in teaching Evan. She began to see the classroom environment as one that did not provide the opportunities for the child to demonstrate what he knew; she began to make changes in the classroom environment and in her practice in order to change that. (For a more extensive presentation of Ellie's case, see Foote, 2009.)

The case of Tina. Tina was attentive to her target student's school mathematical thinking throughout the project. A few weeks after school had begun, Tina reported to the group that her target student, Tyrone (also a kindergartner), recognized most of the numbers between one and ten and that he could rote count up into the teens. She also reported specifically on his (incorrect) appropriation of a strategy during problem solving. In the planning meeting for her second presentation, Tina reported that Tyrone continued to be aware of and experiment with solution strategies used by other children. "Tyrone is really good friends with another child who has some more advanced kind of strategies for solving [problems]. He'll count on and things like that. I see Tyrone experimenting a little bit with some of those."

When meeting with Tyrone's mother to discuss the photographs the mother had taken, Tina learned that Tyrone has an intense interest in video games. Tina recognized that there might be some mathematical thinking involved in playing the video games. On numerous occasions she questioned how she could capitalize on Tyrone's interest in these video games. She speculated that there might be aspects of video games that she could use to support Tyrone's mathematical thinking. She again raised the issue in her poststudy interview when she said, "Obviously, [video games] are very engaging. It makes me wonder about the things they're doing in video games; what ways that video games engage them. Maybe we need to be more engaging the way that video games are." And yet, she made no movement in that direction during the course of the project. Although she has specific knowledge of Tyrone's mathematical skills she did not make a link to out-of-school practices. Tina chose to devote effort not to exploring the potential of video games, but rather to implementing a pull-out model experience for a small group of struggling kindergarten learners. Although this effort built on the model of the study group in that it had several kindergarten teachers working with a struggling student from each of their classrooms, it remained very much based on and oriented by school norms and the development of school mathematics skills and concepts. (For a more extensive description of Tina's in-school study group see Foote, 2008.)

Final Writing

For the final writing, each teacher was asked to write a letter to the next year's teacher of her target student. Five of the six teachers mentioned that the home environment of their target child was a supportive one. This is particularly interesting in light of the fact that at the onset of the study, the teachers' stance toward the home environment was not a positive one. These data, along with comments made in the poststudy interviews, indicate that the majority of teachers had repositioned the home environment away from being an impediment and toward being a support for the child's learning. It seems likely that engaging with the parent and closely examining the target student contributed to this repositioning.

Final Interview: Plans for Action

The poststudy interviews demonstrated that teachers had decided to replace some of their classroom practice because of what they had learned in the course of the study group. They had either implemented new classroom practices or had made plans to do so. A common theme that emerged was that of focusing on individual children. In addition, a common concern for developing a deeper understanding of children's mathematical thinking runs through the comments that teachers made.

In general, teachers indicated that they planned to focus more on individual learners or to connect more strongly with families. One teacher, for example, expressed this by saying that she wanted to build more small-group time into mathematics instruction, and eventually build more one-on-one time into her program. Another teacher focused on building ties with families. She reported that she had plans to make more contacts with families. She saw the value of this from the relationship she built with her target child's mother and wanted to extend this to other families. In a small step toward this goal, she began making more phone calls to parents to discuss their child's learning strengths.

Teachers also implemented plans that went beyond the boundaries of their individual classrooms. For example, as we saw previously in the case of Tina, she encouraged the other teachers at that grade level to devote more attention to issues of mathematics.

Discussion

The goal of this study was to investigate how participation in a study group in which teachers each explored the mathematical thinking and in- and out-of-school experiences of a child from a nondominant background contributed to those teachers' growth and development as teachers of mathematics. I begin this discussion by synthesizing the results for each research question looking at what teachers learned about children as mathematics learners and about their own mathematics teaching practice. I conclude by discussing implications for practice and future research.

Individual Children

Five of the six teachers started to think deeply about or make connections with their target children. From among this group, Ellie stands apart as an exemplar of what is possible. It would have been ideal if all the teachers had been able to learn and apply their learning in the way that Ellie did. She not only identified her target child's specific mathematical abilities outside of the classroom, she also brought that learning into the classroom. She discussed competencies such as Evan's ability with counting that she had accessed while consulting with his mother. She pondered ways to bring that out-of-classroom competency into the mathematics classroom. She not only learned things about Evan, she learned things about herself as well. She moved from dismissing Evan's mother's estimation of his counting abilities to considering more carefully that her own experience was one that should be open to examination. She considered seriously and acted upon information from the parent that was at odds with her own experiences.

The degree to which Ellie, compared with the other teachers, was able to identify already existing classroom structures that both supported and frustrated her target student also sets her apart from other teachers. Through the course of her work in the study group she identified that Evan worked best in the workshop rather than the centers format. This analysis of the characteristics that supported Evan's learning may have allowed her to think about how to restructure other time in the classroom in a way that was not as available to other teachers. This ability to analyze existing practice, and analyze it not merely in a general way, but with regard specifically to her target student, may have positioned her to interrogate her practice in a way that supported her in either implementing or making plans to implement changes in practice.

Tina as well as others learned about their target student's out-of-school interests. The degree to which these teachers recognized that this learning might be significant and might be brought into the mathematics classroom varied. Some, like Tina, reported recognizing that their target student's out-of-school interests had potential for supporting mathematics learning, but they did not build on this learning. Others talked less positively about their target student's interests, although they were able to identify some areas as possibly related to mathematics.

These cases confirm Villegas's (1993 as quoted in Zeichner & Hoeft, 1996) findings that conferring with parents and observing children in school are ways that give teachers access to children's lived experiences. But all the teachers except Ellie had a difficult time building on the mathematical competencies that they had identified in their target child's out-of-school experiences. The difficulty that several teachers experienced in this study in connecting out-of-school practices to school mathematics is consistent with Civil's (1998; 2002) findings that connections between in-school and out-of-school mathematics are difficult to make.

Three of the teachers finished the study group without a plan to connect their target students' out-of-school interests to mathematics instruction, but they did report that they had built strong personal relationships with their target children and with their families. The knowledge they gained about the child and family may have supported both the deepening of the relationship and a view of the home as a resource. All three talked passionately about the strong relationships they had forged with their target students. They said they had made strong positive connections to their students and that this positioned them to be more effective teachers of those children. They said that they felt committed to them and that this commitment was being translated into improved performance in school as well. It may be that these teachers developed (or conceivably already possessed) some important relational knowledge for teaching (Grossman & McDonald, 2008). Perhaps because of the deep personal connections they had developed that seemed to be supporting improved performance in the classroom, they felt less urgency to connect the children's out-of-school experiences to classroom mathematics.

All Children

The teachers initially placed much of the responsibility for failure to be successful in mathematics within the home of the student. These findings are consistent with the findings of the funds-of-knowledge researchers (González et al., 2001; Moll & Greenberg, 1990), who also reported that teachers began their involvement in the funds-of-knowledge projects with deficit views of the homes and communities of their students. In their final writing, five of the six teachers in this study described the homes of their students as supportive. The growth that teachers exhibited in coming to view the homes of their students as supportive environments also confirms the findings of Moll, Civil, and their colleagues, who found that the relationships that teachers built with students and their families challenged the teachers' deficit views of the home environment. This certainly was the case with Nan, who felt that her target student was progressing in mathematics, supported in part by the strong relationship she had formed with the student and the student's family.

In a way that Ladson-Billings (1994a) suggests is important for teachers, several teachers in the study connected more closely with the families and communities of their target students, and then with families of other children in their classrooms. Several teachers (Nan, Meg, and Rebecca) developed more open channels of communication with parents so that parents' concerns and points of view were brought to the table with those of teachers on a more regular basis.

Teaching Practice

Thinking about teaching practice was clearly something that was a central issue for the teachers throughout the study. Although the presentations focused on individual children, the writing that the teachers did at the conclusion of each study group session overwhelmingly focused on issues of practice. At the conclusion of the study, all six teachers discussed having considered anew some aspect of mathematics teaching, attending to one or more instructional issues such as (a) the role the teacher plays, (b) the selection of appropriate tasks for students, and (c) the management of mathematics instruction. In all of these cases, the teachers' consideration of their mathematics teaching practice focused on issues of attending to students' thinking in mathematics. When discussing the role of the teacher, for example, the comments centered around connecting to the individual learner on the basis of an evaluation of his or her needs and with an eye toward validating and building on prior knowledge. In a similar way, the teachers' attention to task selection focused on meeting the needs of individual learners by selecting or constructing tasks that were based on where children were in their developmental trajectories as well as based on children's interests. Even in the case of discussing the management of instruction, teachers reported efforts to develop documentation systems that allowed them to keep track of individual children's specific strengths and needs in mathematics. Four of the six teachers had begun by the end of the study to implement some new instructional plan related to their mathematics teaching that focused on the knowledge of students.

This attention to student thinking, however, was not in most cases broadened to include building on student competencies outside of school. As previously noted, the teachers began the study with school-based notions of mathematics teaching and learning. This view of mathematics did not shift significantly during the study. It seems then that learning about children's lived experiences does not necessarily open new ways for teachers to view mathematics teaching and learning. Teachers' reports of changes in practice in response to their learning during the study group is exhibited most strongly in the case of Ellie who is singular in the degree to which she was able to process information from outside of the classroom and make plans and begin to implement changes to her teaching practice in response to it. Why was Ellie able to connect what she learned about her student to classroom practice in a way that others were not?

There is not an easy or a straightforward answer to this question. Teachers' receptivity to learning can be constrained by many factors. Personal situations can leave them less available for examining their practice or their relationships with students. The view of mathematics as school bound and solutions to issues of mathematics performance as also school bound is another significant factor that is perhaps most clearly seen in the case of Tina. It may have been easier to act within the boundaries of the school setting, establishing a teacher study group as she did, because she was able to access and connect with resources that were located there. It may be as well that taking action within the school sphere was more familiar and therefore easier.

Implications for Practice

The results of this study demonstrate that professional development that focuses teachers' attention on individual learners and their in- and out-of-school experiences can, at least in some cases, support teachers in reflecting on and changing practice. The model may need some adapting for the particular circumstances of accessing and then building on insights from shadowing and learning about children's out-of-school experiences. Although most teachers accessed information about their target children that seemed as if it could have been built on in the mathematics classroom, few were actually able to do so.

This indicates that teachers may need more support than this study group offered in taking up this information and adapting their practices to accommodate it. Focused attention on identifying the relevance of out-of-school practices to mathematics learning may be warranted.

Tasks could be constructed individually or in a group that could then be piloted in classrooms and analyzed afterward. In the case of an activity that children find engaging, but that remains remote to teachers, such as video games, this technology could be brought to a study group where teachers could interact with it and analyze it for possibilities that it might contain for connecting to mathematics instruction.

Implications for Future Research

It may be worthwhile as well to support teachers in analyzing current classroom structures and practices with an eye toward how they support particular children's learning and performance. Incorporating a classroom practices component in a future study might support teachers in identifying structures within their classrooms that support struggling learners. Making the links between learning and practice proved a challenge for many of the teachers. It may be fruitful to study the extent to which additional supports promote the taking up of this learning.

Note

1. Names of all participants, both teachers and children, have been changed to ensure anonymity. Pseudonyms were chosen to mirror the names of the participants, not to suggest that all middle-class White teachers, for example, have a particular class of names.

CHAPTER 4

Pursuing "Diversity" as an Issue of Teaching Practice in Mathematics Teacher Professional Development

ANN RYU EDWARDS

The prevailing discourse about education in the United States is rife with language about diverse classrooms and schools, the challenges of increasing diversity, teaching diverse students, and so on. Diversity is a complex and thorny concept that is critical to the future of mathematics education and has profound implications for how the functions, organization, and effectiveness of schooling are understood more broadly. Yet, what diversity means to educators and for teaching and learning is hard to pin down. Does diversity simply refer to the existence of difference in classrooms or schools? Does diversity signal a concern regarding the impact of culture, race, class, gender, and language on teaching and learning? Is diversity in itself a social good? Through a close examination of the work of grade-level groups in a middle school mathematics professional development (PD) project, a goal of this chapter is to shed light on how diversity is constructed as meaningful to educators by addressing what diversity means to teachers in a particular context and at a particular time and how they raise and pursue issues of teaching related to diversity in the context of professional development.

Professional Development Context

This chapter reports on a study of a university-district professional development (PD) partnership focused on middle school mathematics. The district, a mid-sized district in an urban setting on the West Coast with a racially, linguistically, and socioeconomically diverse student population, sought to address long-standing race- and class-based mathematics achievement gaps in its secondary schools. Two principles guided the partnership[1]: collaboration as an equal partnership between school and university participants, and the importance of learning in and through practice for meaningful professional development (Ball & Cohen, 1999; Cochran-Smith & Lytle, 1999). The participants (about 30 in total) represented a broad array of professional commitments—teaching, administration, research, curriculum development, and professional development. A team consisting of representatives from all participating schools and all middle school grade levels, district administration, university faculty, and graduate students planned and led the activities. A primary locus of the PD activity was a biweekly partnership-wide seminar. The goals of this seminar were emergent and dynamic, reflecting the leadership team's understanding of the needs and concerns of the participants over time.

The study focuses on the partnership's first year. One early goal was to explore the meaning of diversity. Activities included writing about diversity and discussing readings related to diversity, equity, and power in mathematics teaching and learning. We discussed key questions and issues participants had with respect to diversity and how it impacted teaching and schooling more generally. However, participants were largely dissatisfied by these activities, reporting that the discussions were too general and "old hat," and that these kinds of discussions were not allowing them to delve deeply into what diversity meant for their teaching. What then emerged was an effort to use the seminars to help teachers identify and work together on issues relevant to diversity and equity *in their own practice* that emerged from a *collective examination of classroom artifacts*. Thus, the organization of the seminar changed so that the participants could meet in parallel grade-level working groups—one focused on sixth grade issues, and the other focused on eighth grade issues. Early on, the participants felt that in order to make progress on instructional approaches that address differential mathematics achievement, they should first better understand how students—or groups of students—differently performed and participated in their classes. They decided to collectively engage examples of student work

"Diversity": An Issue of Teaching Practice   61

and video recordings of their teaching and of their students working in small groups. Each group was given the latitude to pursue this agenda in ways that reflected their needs and readiness.

The study examines how diversity came to be constituted as a *problem of teaching practice*[2] in the two grade-level working groups. How is diversity conceptualized as relevant to the work of teaching mathematics? What aspects of practice are demonstrated as consequential or problematic in that context? How are those issues of practice engaged and pursued? I found that teachers' conceptualizations of diversity were strongly situated in their local teaching contexts, in particular the historical and social organization of mathematics teaching at their grade levels in the district. Furthermore, their different framings of diversity shaped the nature and status of the practice-related issues that emerged in the inquiry activity as well as the way these issues were collectively pursued.

Data and Methods

Data include audio/video recordings of all working group meetings during the first year (approximately biweekly), field notes from leadership team planning meetings (and related e-mails), participants' writing during the PD seminars (including the group meetings), all artifacts used during the PD seminars, and interviews with selected participants. Data analysis involved three complementary foci—diversity, teaching context, and problem engagement—each contributing to pictures of the two groups' treatment of issues of practice related to diversity and equity, situated in historical, social, and interactional contexts.

The first analytic focus examined how participants talk and write about "diversity" in the various contexts of the PD (including but not limited to the grade-level groups). Through an open coding process, I developed grounded categories of diversity-relevant issues, particularly attending to how the terms "diversity" and "equity" (as well as related variations) were used in context. Second, drawing on observations of and conversations about teaching with participants across multiple settings in the partnership, I examined the histories of and factors that influenced mathematics teaching at these grade levels, paying particular attention to the diversity-relevant issues previously identified. What emerged was an understanding of the conditions and tensions that shaped teaching practice in the district, the particular challenges that manifested at different grade levels, and the locally historical ways in which they had been engaged. Lastly, analysis shifted to how the groups identified and took up problems of practice in their meetings. Analysis of the meeting transcripts was initially thematic; then discussions were parsed into topical chunks to map the thematic flow. A subsequent fine-grained pass coded participants' *noticings,* an analytic unit in which a speaker introduces and articulates an event or issue of note, often invoking evaluations or conclusions (Ryu, 2006) to track how issues emerged and were taken up. The noticings analysis in conjunction with the episode mapping revealed differences in the patterns of problem identification and engagement in the two groups.

For each of the two cases I describe the historical and institutional context of teaching in the district relevant for each of the grade levels, the manner in which diversity was framed by the groups as a problem relevant and consequential to their practice, and the manner in which each group designed and enacted ways to pursue and address that problem in the PD. Due to limitations of space, only the eighth grade case is presented in depth. A sketch of the sixth grade group is provided to highlight contrasts in the groups' foci and processes of inquiry into diversity.

Eighth Grade Group: Making Sense of Diversity as the Challenge of Algebra 1A

The eighth grade group organized their inquiry activity to address a critical and collective issue of practice—the teaching challenge of Algebra 1A, a low-track, eighth grade algebra course. They designed their investigation to reveal ways in which Algebra 1A students' mathematical problem solving differed from that of Honors Algebra students, in order to understand how to better serve the needs of 1A students. Through the work of specifying their students, hypothesizing explanations for what they observed, contesting hypotheses, and constructing foci for further investigation, the group was able to make progress on understanding and addressing the impact of diversity on their teaching.

Transitioning to "Algebra for All" in Eighth Grade Mathematics and the "Challenge" of Algebra 1A

The primary concern of the eighth grade teachers during this year of the study was the new organization of the eighth grade mathematics courses in the district. As part of a new "Algebra for all" policy, the district had reorganized the middle school mathematics offerings so that all eighth graders were enrolled in an algebra course. The goal was to ensure that all students have access to college preparatory mathematics in high school, which made enrolling in Algebra 1 in eighth grade a necessity. Historically, the middle schools in the district preferred untracked, grade-level math courses; however, in recent years both the seventh and eighth grades offered multiple tracks in response to community pressure.[3] Just prior to the "Algebra for all" initiative, the eighth graders in the district were placed into one of three tracks: "eighth grade math," "regular" Algebra, or Honors Algebra. In this year, a new course, Algebra 1A, was introduced into the district middle schools, replacing the "eighth grade math" track. Algebra 1A uses the same curriculum as the other algebra courses, College Preparatory Math (Sallee, Kysh, Kasaimatis, & Hoey, 2002), though implemented at half pace, with less emphasis on collaborative group work (an integral part of the CPM curriculum) and supplemented with more remediation and less enrichment.

The decision to enact this curricular change was made only a few months before the start of the fall semester. Initially, despite the lack of forewarning and preparation time, most of the middle school math teachers (and importantly, almost all of those who taught eighth grade) believed the curricular reorganization toward "Algebra for all" to be well intentioned, and were enthusiastic to take on the challenge. Indeed, at the two middle schools with more than one eighth grade mathematics teacher, the most senior and well-respected mathematics teachers volunteered to teach this new course. The same teacher taught all of the Algebra 1A courses in each of the three middle schools. These three teachers quickly found teaching Algebra 1A to be extremely challenging. They had had limited opportunities to discuss the curriculum and their planning prior to the start of the school year, and so they took advantage of the PD working group to share and compare their experiences. They coordinated pacing and assessments and shared some of the difficulties they were facing, including classroom management issues, problems keeping students engaged, and doing the necessary remediation while keeping the content algebraic. All three were also teaching Honors Algebra. They noticed significant and troubling differences in the problem-solving competencies and effectiveness of collaboration between their Honors Algebra students and their 1A students. They began to wonder, along with their colleagues in the group, what they could learn about effectively teaching algebra to these different groups of students by better understanding their mathematical and social behaviors.

Additionally, many participants of the professional development partnership more broadly were concerned about the equity implications of the new algebra tracking structure. In particular, the racial and socioeconomic divide between the tracks was often cited as a "diversity issue" in the seminar. For example, in an early writing exercise, an eighth grade teacher cites a disproportionate range of racial and socioeconomic representation in low-track classes as the "biggest diversity-related challenge" he faces:

> At [my school], when you look at classes of students behind in math, African American students are apt to be over represented. By contrast, in our honors level courses white and Asian students are over represented. Of course, behind these patterns, socioeconomic division in this town just happens to favor those groups that benefit from our honors classes. Everyone decries this state of affairs, but every attempt has essentially focused on improving teaching—which tends to raise all ships equally. Perhaps this is all we can hope for? Hopefully not.

In the eighth grade, "classes of students behind in math" are the Algebra 1A classes. Thus, what this teacher is pointing to is the challenge of teaching classes like Algebra 1A in ways that "close the gap." Rather than just generically improve teaching, he wants to better understand what one of his eighth grade colleagues called "what best serves 1A students."

In some ways the situation in this district's middle school mathematics courses reflects well-documented trends: Lower tracks have disproportionate numbers of African American, Latino, English Language Learners, and low-SES students, and content and pacing expectations are less demanding (e.g., Oakes et al., 2004). However, contrary to what is found in most schooling environments (Talbert & Ennis, 1990), the lower tracks in the eighth grade were taught by high-quality, respected teachers who had a stated "preference" for heterogeneous, untracked classrooms. When faced with the teaching difficulties in Algebra 1A together with the troubling and enduring racial and ethnic profile of their low-performing students, they recognized the "Algebra 1A situation" as a problem related to their teaching practice and chose to take it up as the object of inquiry in their group.

Eighth Grade Group Inquiry Activity: Making Sense of "What Best Serves 1A"

In much of the work reported in mathematics education on the use in PD of artifacts of student problem solving, the aim is to unpack students' mathematical thinking in order to learn about the development of mathematical understandings, to develop instructional strategies for revealing and probing students' thinking, and to come to understand the diversity of ways in which students engage mathematical content (Hatfield & Bitter, 1994; Kazemi & Franke, 2004; Sherin & van Es, 2005). The eighth grade group, however, chose to use artifacts (video recordings and written work) of student problem solving to identify and unpack the needs of a targeted population—the Algebra 1A students. They hoped that this comparative examination of their students' problem solving would help them identify the particular difficulties that 1A students faced and begin to shed light on how they could better address those difficulties and thereby "close the gap" that they were so concerned about. Specifically, the Algebra 1A teachers, with the help of university participants, videotaped groups of 1A and Honors students working on the same or similar algebra word problems. From these recordings, the group chose to view four clips of students solving problems in small groups: Matt's Honors class, Matt's 1A class, Dave's 1A class, and Dave's Honors class[4]. The cross-case comparative structure of the inquiry activity set the group up to attend to differences between 1A and Honors students as well as similarities between students at different schools and with different teachers.

In what follows, I sketch the arc of the eighth grade group's work, focusing on three critical phases of their inquiry: (a) specifying 1A students as doers and learners of mathematics, (b) constructing and debating hypotheses of "what best serves 1A students," and (c) negotiating the purposes and foci of further inquiry.

Specifying 1A students as doers and learners of mathematics. In the early conversations, two primary themes emerged: (a) how 1A students do and learn mathematics differently from the Honors students, and (b) how 1A and Honors students behave in interaction with their peers, as contributors to and participants in learning. The participants described 1A and Honors students in much the same ways that studies have found teachers characterize low- and high-performing students (Reed, 2006; Stodolsky & Grossman, 2000). Broadly speaking, the group characterized the mathematical knowledge, behavior, and skills of the 1A students as weak: They have poor calculational skills, exhibit serious problems modeling and interpreting mathematical situations, and have trouble communicating mathematically. In contrast, the group found that the Honors students in the videos possess a greater facility with mathematical relations and representations, are better explainers, and are better able to think analytically about mathematical situations. They also concluded that that the Honors students engaged in more "fruitful" collaboration with their peers, taking responsibility for one another's learning, while the 1A students, though perhaps more "active," were less "productive" and more "confrontational."

If left at this level of generality, the group's conversations may not have provided openings for unpacking and learning about their 1A students in ways conducive to rethinking practice. However, unlike most talk among teachers about students in tracked classrooms (Horn, 2007; Stodolsky & Grossman, 2000), this group sought to discover and ground observations of the mathematical thinking and behavior in the details of actual classroom events. Thus, their specification (Horn, 2007) of these "kinds of students" developed as the group's discussion moved between close examination of the students' written work and interactions in the clips, interpretations of students' thinking and behavior, and examination of possible reasons for the differences they were seeing. They maintained a focus on those aspects of students' mathematical experiences over which they felt they had some influence—teaching practice and the organization of the district curriculum. In this way, what could have been an unproductive and damaging exercise in labeling student deficits instead resulted in the beginnings of inquiry into teaching practices to address students' specific needs.

Debating "What Best Serves" 1A students. This attention to the impact of instruction on the needs of 1A students led, over time, to an exploration of how they are served in homogeneous, tracked environments versus heterogeneous, untracked environments. The mathematics departments in the middle schools historically had a preference for heterogeneous classrooms serving a range of ability levels. Teachers and school administrators initially viewed the tracked algebra situation as "hopefully temporary"; however, district policy was as yet undecided regarding the future of algebra tracking, and the existence of Algebra 1A in particular. Against this context, after several discussions, a participant asked, "What do you think would have happened if the Algebra 1A students had been put together with the regular algebra kids?"

The resounding response from the 1A teachers was that they would have been less successful because the heterogeneous environment endangers their confidence and hence their engagement. As the

"Diversity": An Issue of Teaching Practice 67

conversation continued, what emerged was a kind of theory of "what best serves 1A students." Though not universally shared within the group, the 1A teachers (at this point) felt that 1A students (and others like them) are not confident in their mathematics and in themselves as learners of mathematics and hence do not participate in heterogeneous math classes. They were concerned that heterogeneous classes can create barriers for some students' participation. Many of these students lack the focus and skills to be successful in heterogeneous classes; thus, they need a "safe" and structured place that shields them from ridicule and social discomfort, allowing them to participate in the mathematics, and thereby providing successful experiences in which they can learn.

However, despite their concern that a heterogeneous setting might disadvantage 1A students, these teachers were uncomfortable concluding that tracking best serves them. They did not see themselves or their departments as advocates for tracking and were surprised to now find that homogeneity has its advantages. As one 1A teacher noted, "I didn't expect to find that 1A works. I mean it's a hard class, really hard, but in some ways it seems better for them." They worried that their observations about Algebra 1A might be seen as a broad endorsement for tracking, and that tracking can create and perpetuate inequalities as some teachers may have perceptions of "low" and "high" kids that exacerbate disparities. As the conversations continued, members of the group began to question and challenge whether homogeneous environments are beneficial for students in the *long run*, opening up assumptions about 1A students as learners and turning their attention to 1A students' educational trajectories.

What developed were two competing hypotheses about 1A students' needs and the ways to address them: a *reactive* view of 1A students that suggests that the "distracted" and "confrontational" behavior of the 1A students as well as their "lack of confidence" are characteristics that need to be managed through specific kinds of classroom structure and disciplinary strategies and a *developmental* view that suggests that in appropriate environments 1A students can learn to participate in productive ways and that, in time, those changes can be capitalized upon in order to facilitate success in higher-level courses. The reactive view suggests that 1A students are best served in homogeneous classrooms in which these structures and strategies can be developed and implemented to meet their specific needs. The developmental view suggests that while homogeneous environments, if designed to support meaningful student participation, can help students to develop the knowledge, skills, and most importantly, the confidence to participate productively, the *goal* of such instruction should be to prepare them for further courses that may not be so designed (in high school and beyond).

The debate over these ideas—which occurred over multiple meetings and included contributions from nearly all of the participants (e.g., preservice teachers, new and experienced teachers, administration, university faculty, and doctoral students)—led to a refinement of a hypothesis of 1A students' learning linked to a learning environment that resembled the developmental view. As Don, one of the 1A teachers put it,

> Well, I mean as Dave as Dave was saying I think the key, it's almost a quantum jump, I think that once a math student has experienced a little bit of success a little of what can happen if you have your skills in order and your times tables, and you can tackle a problem and eventually solve it, once you get to that point and once you have the confidence to, well, even if it's not solvable I can struggle on it, it's not going to come back down as a heavy sort of emotional difficulty that I'm a failure I've always failed I'm not going to succeed at this. My point I think is that if in the seventh and eighth grade in this sort of protected environment can allow a student to achieve a certain amount of success at it, just that feeling that aha feeling and the craving for repeated aha feelings will push that kid forward in a heterogeneous math class.

Negotiating further inquiry. In the last phase of their work on this issue, the group decided to design a subsequent investigation of this hypothesis. There were significant differences between participants as to the ultimate goals and the design of such an inquiry. Some felt they should investigate the advantages and disadvantages of eighth grade algebra tracking by comparing instruction and student engagement in heterogeneous and homogeneous classrooms. For example, Dave argued that pursuing their inquiry into tracking this way "forces the issue" of diversity more directly into their investigation. He advocated that they keep in mind that their original goal was to understand "what teaching diversity is":

> Another idea, I'd caution us not to— I'd like not to just sort of, you know, a task, like you know, let's figure out how to measure whether algebra 1A would be better folded in. I think, you know, we started looking at these tapes at unpacking what it meant to teach you know to different kids, and what teaching diversity is,

and I'm still unclear on what diversity means. I mean it's easy to say it's a matter of race, or it's a matter of skill, but I think it's much richer than that, I think.

He continued to argue for thinking about what teaching methods foster learning in diverse classrooms, and in particular, for investigating their own teaching practices and decisions *analytically* as they engage the new clips:

> Continuing to just talk about what, you know, going back to what you were saying which is, what are good methods of teaching with a diversity of kids, whether it's a heterogeneous classroom or a homogeneous classroom. You know, what works and what is teaching in these classrooms all about, and what does it look like. Because I know what it looks like through my eyes, but that's sort of—you know, it's sort of, it is completely, what's the word, it's not very analytical, it's very, it's just how I teach, you know. So I don't really know what I'm doing. And I'd like to figure out what I'm doing, and then start talking about, you know, do we want to get rid of a course or not. But I don't think we should get too sidetracked by that, you know.

In contrast, others, concerned with issuing a recommendation to the district about algebra tracking, saw this as an opportunity to compare heterogeneous and homogeneous classrooms as a way of understanding the relationships between type of student (e.g., 1A, Honors) and composition of classroom (heterogeneous, homogeneous). Still others were interested in better understanding the reasons for the racial and socioeconomic divide they saw in the eighth grade tracks. They suggested looking at "what happens in earlier grades" to low-performing students—in particular, students of color—and also thinking about proven strategies for effectively teaching in heterogeneous classrooms.

In the end, they compromised on a study that looked comparatively and systematically at students in heterogeneous and homogeneous classrooms across grade levels in order to better understand the teaching practices that shape any differences that they might see. Unfortunately, due to resource nonavailability and work circumstances, they were unable to fully implement their plan. Over the course of the year, however, the eighth grade group managed to collectively identify and pursue a problem of practice related to diversity that was consequential for their instructional decisions and practices. For them, the district's decision to track algebra in the eighth grade created not only a serious instructional challenge, but also an opportunity to closely examine their students and the ways their teaching practices and the structures of schooling impacted those students. Although their work did not result in changes in district policy, over the following three years, the teachers (in collaboration with the university participants) developed an approach to equitable teaching based upon Complex Instruction (Cohen & Lotan, 1997a) and other strategies for promoting equitable participation.

Sixth Grade Group: Diversity as a Problem of Managing Heterogeneity

What follows is a brief sketch of the context of sixth grade teaching in the district and the work of the group in the PD meetings. Sixth grade classrooms in the district were untracked and more reflective of heterogeneous elementary school classrooms than discipline-based middle school curricular organizations. Many of the sixth grade teachers reported that the "greatest diversity-related challenge" that they faced in their teaching involved successfully meeting the needs of the diverse population of students in their classrooms. For example, Ruth, a teacher with over 15 years of experience, expressed concern about "meeting the needs of all my students, especially the academic needs" and wondered how, as a teacher, she could "figure out what specifically is holding a particular student back (be it academic, social, cultural, emotional) and then how do you move that students forward in their mathematical understanding?" She saw her work as determining the needs of any particular student and finding ways of addressing those particular needs. Indeed, throughout the course of the year, she often spoke of how she saw her students as individual people with individual histories and difficulties. She felt that her teaching needed to be tailored to address the complexity of the mathematical, social, and emotional character of the group of *individuals* that constituted her classroom.

The sixth grade math teachers faced a relatively more stable teaching context than the eighth grade teachers. Historically, the sixth grade was positioned as a transitional year between the more nurturing environment of elementary school and the more independent and academically challenging upper middle school grades. Many of the teachers taught multiple subjects to the same students, and nearly all of them were originally credentialed in multiple subject elementary education and had elementary teaching experience. They brought a concern for the whole child and shared a belief that they should prepare students not only for pre-algebra and algebra content, but also provide them with the social and academic skills necessary to meet the demands of middle school and beyond. Individualized instruction was a major focus for the sixth grade teachers in the study. As one teacher noted, "We try to see our kids as individuals and tailor what we do for them as individuals."

Sixth grade math teaching was historically highly decentralized in the district; most of the teachers seemed to do their own thing even though many of the same curricular resources were available throughout the district. The teachers placed high value on their autonomy, and many resisted movement toward common, externally mandated methods or curriculum. They reported that the freedom to tailor their teaching was of paramount importance. Thus, pedagogical approach, curriculum, and pacing varied greatly among the sixth grade math classrooms. This autonomy was reflected in their PD meetings, in which sharing and comparing, in contrast to in-depth analysis and questioning, was the norm. Teachers were respectful of one another's expertise and understandings of their own students and were often unwilling to come across as critiquing their colleagues. Thus, while the prevailing autonomy in sixth grade mathematics teaching allowed teachers to pursue their own vision and home in on the specific needs of their students, this norm seemed to limit the opportunities to critically engage issues of practice in the PD.

The group decided to use the PD inquiry activity as an opportunity to use available artifacts of practice to analyze student mathematical thinking and seed a conversation about teaching practices that would address how to "reach all students." Over the course of several meetings, the group looked at examples of student written work as well as videos of teaching.

Participants noted interesting, puzzling, or problematic issues stemming from these artifacts that triggered discussions of student thinking—especially common misconceptions—and the sharing and comparing of instructional strategies and classroom experiences. This sharing included how to address differences in student participation, how to implement rich tasks equitably without penalizing those with fewer supports and resources, and how to foster meaningful mathematical explanations.

However, the central issue of "how to reach all students" was not treated in a coherent manner. Some participants reported that the conversations felt disjointed and lacked focus: "There were just lots and lots of pieces, I mean I don't think we synthesized them or ordered them in a list, and we certainly didn't get to making an agenda for what should come next." Also, members of the group expressed frustration with the paucity of talk about diversity in their conversations. Jim, a veteran sixth grade mathematics and science teacher, reflected,

> I enjoy this conversation, I think that it's extremely rich, but sometimes I wonder how much of this discussion could be going on in a thoughtful, professional group in North Dakota where there is less diversity. I mean a lot of this is just, math teaching and pedagogy and good things to think about, but I don't know how much of it addresses, for me, dealing with diversity of kids. So I don't know how—I keep at the end of each session wanting—and this, is this going to help you reach you know a greater range than I'm currently reaching.... I don't know if it necessarily attacks the kids that I'm not succeeding with.

What Diversity Means to Teachers and Teaching

These two cases show that the ways in which diversity is understood as a problem of practice for these groups were strongly shaped by the local institutional and political histories of teaching at different grade levels. Mathematics classes in the sixth grade were untracked and seen as a transition from elementary to secondary education, and the sixth grade math teachers highly valued individualized instruction. The sixth grade group framed diversity as a problem of heterogeneity, that is, how to justly address the variation in a heterogeneous classroom by successfully reaching all students with appropriate and substantive mathematics content. In contrast, the focal issue of diversity for the eighth grade group was the challenge of teaching the Algebra 1A course, a new low-track algebra course chiefly for low-performing eighth graders created in the wake of a new "Algebra for all" mandate in the district. Their problem was one of homogeneity—how to successfully teach algebra to homogeneous groupings of students in different tracks.

Furthermore, the approaches of the two groups to investigating these issues also reflected the contexts and practices of teaching in the two grade levels. The sixth grade teachers across the district were highly autonomous and individualized in their curricular and teaching practices. Although they valued the opportunities to share and compare their experiences and their expertise, they did not organize or engage in their inquiry in a way that supported pedagogical sense making of a collective issue. Each artifact triggered multiple and different issues that were thematically related but did not build toward broader insight into the key challenge of reaching all students. It is likely that stronger facilitation on the part of a member of the leadership team might have made a difference for the depth and coherence of the group's discussions over time; however, this might have been bought at the cost of teachers' willingness to participate and the building of an even minimal comfort zone for discussing one another's practice.

In contrast, the teachers in the eighth grade group were all facing a new and challenging teaching context—tracked algebra courses. They felt that the working group could be an opportunity to better understand and address the difficulties that they felt, and at the same time, better understand and address the ongoing disparities in achievement and participation among their students. The sense of shared experience and focus as well as the immediacy of the issue were significant for how the group organized and pursued their inquiry. Although their discussions were rife with disagreement and debate, there was a clear collective investment in making progress. This is not to suggest that the outcome of the group's work in this first year was somehow exemplary; indeed, some members were disappointed in what they saw as a lack of concrete results. However, the conversations of the group did lay the groundwork for subsequent sustained work in lesson study and complex instruction.

Conclusions and Reflections

These cases underscore the argument that professional development efforts must be informed by and grounded in the actual lived contexts of teachers' work. The meanings that people make of the activities they are engaged in and the issues that emerge in those activities, as well as the orientations that they bring to the collective work of thinking about practice, are shaped by the day-to-day situations and challenges they face in the classroom. Of course, this principle for professional development is not new (e.g., Cochran-Smith & Lytle, 1999); however, this study highlights the vastly different understandings that people can arrive at regarding something that everyone agrees is a centrally important issue—diversity, for example—, as well as the implications of those differences for the nature of their inquiry. Too often, university-based professional development efforts—well intentioned and informed by research—pursue their own agendas with little consideration for the specifics of local histories. This is particularly true of PD addressing issues of diversity and equity. Because the needs are so immediate and the injustices so apparent (to us), the temptation can be to tell teachers what to believe and what to do rather than listen to what they think, observe what they do, and examine *why*. Just as current thought about student learning has embraced the notion that students bring knowledge, beliefs, values, and practices into classrooms and that teachers are well served not only to be aware of them but also to leverage them in their teaching, those designing and offering professional development should examine participating teachers' lived experiences and capitalize on those understandings in creating learning experiences (Bransford, Brown & Cocking, 2000).

This may be particularly important for professional development that requires explicitly examining sometimes difficult beliefs about students, teachers, teaching, schooling, and our own roles in reproducing and resisting inequity. The beliefs that frame teachers'—as well as our own—interpretations of the achievement gap, and in particular, perceptions of low performance, disruptive and confrontational behavior, lack of motivation, and so forth, of poor students and students of color, are deeply rooted in lifetimes' worth of experiences and the messages about race, culture, and class that we continuously negotiate in U.S. society. Critical examination of these beliefs is a formidable task; *changing* beliefs and teaching practice is even more so. The PD efforts described in this chapter were not wholly successful in achieving these goals, but some lessons were learned. Opening up one's beliefs to what is perceived to be public scrutiny—indeed, opening up one's *identity* in important ways—depends on a mutuality of respect and trust that can take months if not years of sustained collaboration to develop. University participants should come to understand—and importantly, become invested in—the daily practice of teachers and the dynamics of school and district politics. Teachers not only need to be given voice and their expertise and knowledge valued, but they need to take (and be given) responsibility for the work. Finally, all need to be held accountable to the students—giving voice not only to their "needs" but to who they are, what they believe, and what their experiences are as mathematics learners.

Notes

1. I do not mean to suggest that the participants were consistently successful at living up to these principles; however, they guided the inevitable negotiations and compromises.

2. By "problem of teaching practice" I do not index negative connotations of the "problem." The use of "problem of practice" relates to the notion of "problematic" in that it presents a situation not easily answered or understood and requires reasoning beyond "easy solutions" (e.g., Lampert's [2003] notion of problem of practice and "problematic tasks" in literature on mathematical problem solving.)
3. Teachers in the study reported that more advanced courses were created in response to community pressure, which, in effect, caused the original courses to become lower tracks.
4. Names of participants have been changed to preserve anonymity.

CHAPTER 5

Teacher Positioning and Equitable Mathematics Pedagogy

ANITA A. WAGER

In this chapter I present findings from a study of a semester-long professional development seminar in which teachers critically examined equitable mathematics pedagogy (the interrelation between teaching for understanding, students' out-of-school mathematical knowledge, and the mathematical knowledge students need to question inequities in their world). In analyzing data generated through the seminar, evidence emerged describing how teachers' histories and previous experiences influenced their trajectories of engagement with equitable mathematics pedagogy. I first provide a brief context for the seminar. Second, I offer a way to position teachers' identities with regard to equity and mathematics. Third, I share case studies of three teachers as they examine their beliefs and practices regarding equitable mathematics pedagogy. Finally, I explore both theoretical and practical implications of the study and how I positioned myself in this research.

The Context: Professional Development in Equitable Mathematics Pedagogy

In the spring of 2007, I facilitated a semester-long professional development seminar in which 17 teachers examined their beliefs and practices with regard to equitable mathematics pedagogy. The seminar was structured to use Ladson-Billings (1994a, 2001) culturally relevant pedagogy as a lens for broadening views of teaching and learning mathematics. The teachers used Ladson-Billings's indicators of academic achievement, cultural competence, and sociopolitical consciousness as prompts to reflect, discuss, and interview each other about their views of teaching mathematics. As a part of a larger case study of the professional development, I used nested cases of three individual teachers to document the different paths some teachers follow as they broaden their understanding of equitable mathematics pedagogy. The teachers were selected to represent different levels of initial understandings and change in practice.

The instruments in the study included audiotapes of seminar discussions, audiotapes of teachers interviewing each other, teachers' multicultural and mathematical autobiographies, reflective responses to readings, pre- and postseminar interviews, and field notes from classroom observations. In analyzing the trajectories of the three teachers, I looked at an evolution of the teachers' ideas about and understanding of the constructs and indicators of culturally relevant pedagogy (Ladson-Billings, 2001) as I had refined them to connect specifically to mathematics. These were countable events that occurred during teachers' conversations. I then triangulated the data with what teachers wrote in their reflective responses, what they said in interviews, and what I observed in their classrooms. Their discussions and interviews with each other and me as well as their written reflections and autobiographies provided the foundation to describe their positioning with regard to this work as well as their trajectories through it.

Identity with Regard to Equity and Mathematics

The teachers in the seminar had a broad range of backgrounds with regard to equity and diversity. Experiences ranged from growing up in privileged White middle-class areas with limited exposure to diversity to encountering moral outrage against racism at an early age. On the basis of teachers' autobiographies, narratives, and initial interviews I have categorized them into three broad groups relative to equity experiences: limited exposure to diversity, immersion in another culture, and exposure to racism. I then examined teachers' views of equity and mathematics and found three main perspectives: never considered equity in mathematics, thought a lot about equity but not as related to mathematics, and already considered mathematics through an equity lens. In Table 5.1, I offer a rough approximation of where the three

Equitable Mathematics Pedagogy

Table 5.1 Teachers' initial perspectives on Equity and Mathematics[a]

Mathematics	Equity		
	Limited exposure to diversity	Immersed in another culture	Experienced racism
Standards-based mathematics enough; never considered equity in mathematics	Group A Rose		
Thought about equity, but not as related to mathematics		Group B Inga	
Already considered mathematics through an equity lens and wanted to learn more			Group C Caroline

Note: [a] These are based on autobiographies, initial reflections, and initial interviews

teachers[1] would be placed in terms of their experiences and initial perspectives on equity and mathematics. I am not using the categories to make strong claims about the teachers but as a general overview of the levels of experiences teachers have with different cultures and racism. Certainly, other factors influence teachers' positioning relative to equity and mathematics. These factors include but are not limited to professional experiences in terms of professional development on equity, teaching mathematics for understanding, and teaching experiences ranging from several years in diverse settings to just one.

Three Teachers, Three Stories

These findings are presented in narrative form to reflect the stories of Rose, Inga, and Caroline as they engaged with the constructs of culturally relevant pedagogy in mathematics. In each case I provide a brief description of the teacher's initial positioning and then describe her trajectory with regard to academic achievement, cultural competency, and social justice.

Rose's Story: Baby Steps

Rose had limited exposure to racism and her initial conception was that standards-based mathematics was enough to achieve equity. After teaching with traditional mathematics curricula for 15 years in another state, Rose moved to her current position and was introduced for the first time to standards-based (reform) mathematics. She embraced the challenge and enrolled in courses, participated in professional development programs, and completed a year-long mentoring program to develop her skills in teaching mathematics with understanding. The success of Rose's conversion was evidenced by her role as head of her school's mathematics action school improvement program committee, her work with the district in developing a new intermediate grades teaching manual for mathematics, and her position as demonstration teacher for the local university's elementary mathematics methods courses. The depth of her belief in the power of teaching for understanding was evidenced by these roles as well as comments such as, "Today, I am a firm believer in its [standards-based mathematics] benefits for all our students.... It is now very meaningful and enjoyable to me" (Autobiography, 2/5/07). The teachers in the seminar quickly recognized her strengths and sought her counsel in matters such as heterogeneous grouping and pedagogical approaches. Rose's story begins with her assumptions that standards-based mathematics is enough to achieve equity. Next, an understanding of cultural competency unfolds, and finally, she questions her initial conception by recognizing that standards-based mathematics is not enough unless social justice is considered.

Academic achievement: Is standards-based mathematics enough? Rose's own exploration in the seminar was distinctly linked to her focus on teaching mathematics for understanding. She felt that standards-based mathematics was enough to address historical inequities. During her first reflective reading response, she expressed this sentiment: "If we look at our standards, and teachers, and the pedagogy now provided for teachers surrounding math, aren't we achieving social justice?" (Reflection, 2/5/07). Rose felt that the effects of using standard-based mathematics had not had time to take hold and that older students had missed the early intervention but over time reform teaching would help close the achievement gap.

Cultural competency: An understanding unfolds. Rose's initial conception was that standards-based mathematics should be enough to provide all students with the opportunity to achieve. Yet, by the third meeting she had already moved toward explicitly considering equity in teaching mathematics. "For me, its keeping equity, math, and the two combined on the forefront of my brain and causing reflection" (Reflection, 3/5/07). Rose built on her strength in academic achievement by extending her knowledge of students' mathematical understandings to also consider the cultural contexts in which students live and learn. She started first by recognizing the need to consider students' cultural experiences when teaching mathematics. In one of her reading reflections, Rose wrote, "Kids need to see themselves and their lives reflected in literature. Certainly, they need to see themselves reflected in mathematics" (Reflection, 4/16/07).

As the seminar progressed Rose took this further by identifying ways in which to incorporate students lived experiences within the classroom. During an interview with another teacher, she pointed out that "we have been doing this in literacy and social studies for some time but, now we are talking about it and you are making me [do] culturally relevant math problems" (Teacher Interview, 4/30/07). Rose had the general notion that problem solving should incorporate contexts that reflect students' different experiences. Rose also took up this notion in her classroom but struggled with the time required to do so in a way that individual students' experiences were used. Instead, she identified experiences that all her children shared, such as using money. As related to cultural competency, Rose broadened her conceptions of teaching for understanding and was starting to consider what this would look like in her practice.

Social justice: Perhaps standards-based mathematics alone is not enough. After reading Gutstein (2006), Rose wrote, "I did like his thought that math and social justice are dialectically interrelated—neither is sufficient by itself" (Reflection, 5/21/07). As with cultural competence, Rose's conceptions about the importance of social justice were changing, but she was only just beginning to try it out in her practice. She drew clear lines about the types of lessons she was willing to do. For example, after reading the chapter about comparing spending on the war to spending on education in *Rethinking Mathematics* (Gutstein & Peterson, 2005), Rose wrote that although the article was interesting, "there is NO WAY I could have math lessons like that. It's a political bomb (pun intended). I'd have administration and probably a few parents all over me! Real life, true, relevant math—yes. But, not acceptable if I want to keep my job!" (Reflection, 5/7/07).

Although Rose had no intention of pursuing a lesson as politically charged as the Iraq War, she did begin to incorporate sociopolitical discussions in mathematics. During a visit to her class, I observed that she started class by sharing a graph from the newspaper that made her nervous. The graph displayed the probability of being in a car accident based on your age. Rose had a son who had just received his driver's license and she asked her students why this graph was important to her and how it was important to them. There was discussion about what the graph was saying, how to interpret it, and what it meant for the students, particularly those whose siblings had recently started driving. In discussing this graph, Rose explicitly showed students the power they had to use mathematics to interpret what they saw. The context had immediate relevance for some students (those with family members who are just starting to drive) but held interest for all. The lesson opener was a first step in teaching mathematics *about* and *for* social justice as she discussed a social issue relevant to students' lives (Field Notes, 6/4/07).

Baby steps. Ultimately, Rose had significant change in her conceptions of what mathematics classes should look like, but acknowledged that the change in practice would take time. "It is just layers. It is like reading Gloria's [Ladson-Billings] book and trying one little piece. It is like being in the class and taking some problems home with you. It just doesn't happen overnight" (Teacher Interview, 4/30/07). Rose welcomed the changes she was making in her practice but referred to that change as taking baby steps.

Inga's Story: A Fork in the Road

Inga's experience with other cultures came from her school setting that had for many years focused explicitly on learning about and building on students' cultures. Her initial conception was that culture was important, but she did not see where it fit in with mathematics. Although she brought her experiences of teaching in a very culturally aware setting, Inga was the most dubious with regard to considering mathematics through an equity lens. In that respect, Inga made the most progress of the three teachers. Whereas Rose assumed cultural and social justice concerns were embedded in standards-based mathematics, Inga did not think they were an issue in mathematics at all. She clearly saw them in other subjects, but like many, considered mathematics to be culture free. By her own admission, she chose to take the class to see if she could be proven wrong: "When I signed up for the seminar, I thought how are they going to get math with that? And then I saw it where I didn't even know it was there" (Postinterview, 6/5/07). Inga's story begins with her perspective that flexible grouping is okay. Next, she starts to see how she can extend her cultural competency to other cultures. Finally, she has a breakthrough as she considers including social justice as a new way to think about mathematics.

Academic achievement: Flexible grouping is okay. A common theme throughout Inga's reflections and discussions of academic achievement was how students should be grouped to provide the best support for all. This topic weighed heavily on her as a result of school policies that influenced her mathematics classroom. At the beginning of that school year the principal informed teachers that they should group students for mathematics on the basis of the previous year's state assessment scores. Initially Inga questioned the idea of grouping students this way but over time she came to agree with the strategy, believing it was best for students. In her first reflection, she wrote, "On the one hand I was very upset about being forced...into this way of breaking apart the kids (especially the use of [a standardized assessment]), but on the other hand it is working" (Reflection, 2/5/07). Here, she is referring to the frustration she felt at the beginning of the school year when students were split in different classrooms for mathematics on the basis of test scores. Despite the resulting homogeneous ethnic composition of the classrooms, Inga came to believe that the groupings worked for the students.

With the exception of classroom grouping, something she had no control over, Inga was open to trying new things. She engaged with the ideas she read about and heard in the seminar and tried them in her class. In reflecting on Guberman's (2004) study comparing out-of-school practices of Latino/a and Korean families, Inga observed that many of her Latino students showed strengths in money and geometry units. "The article has helped me to think in terms of what success my Latin American kids are having because of what many Latin American families encourage at home" (Reflection, 3/5/07).

Later, Inga took up the ideas prompted by the article and the responses she received from surveys about out-of-school mathematics her students had completed. In noting that some students shopped for their families, she decided to interview them about their experiences with money. From this, Inga learned about her students in ways she didn't expect.

> Kids who I didn't suspect knew money, knew it, down cold. Kids that I thought would know it, no problem, didn't know it at all or just knew the values of coins and didn't know how to make change.... One of my boys, [Jamal], he does all the shopping for the family. He walks down to the PDQ and gets all their food and he knows money cold. And he is just very average, very average. But he knows money the best of any kid in my class. (Postinterview, 6/5/07)

This was a breakthrough in many ways for Inga: (a) she recognized that her preconceived notions about a student were wrong, (b) she identified mathematical strengths in a student that she was unaware of, and (c) she realized that students' out-of-school mathematical practices provided a foundation upon which to build. Although this was a breakthrough for Inga, she could have taken this further by exploring the specific strategies Jamal was using. The strategies children use with money are often nonroutine and this might have offered an opportunity to gain a deeper knowledge of Jamal's understanding.

Cultural competency: Extending it to other cultures. Inga was considered by others at her school as one of the most culturally competent with respect to African American students. According to one teacher in her school, parents often assumed Inga was bi-racial because of her pedagogical practices, not the way she looked. She also felt competent in understanding her African American students but worried that she had ignored her Latino/a and Hmong students. In a reflection, she wrote, "For years I have been immersing myself in the lives, interests, etc. of my African American families but have had little outside time with my Latino and Hmong families" (Reflection, 3/26/07). During the semester Inga became increasingly interested in broadening her understandings of Latino and Hmong cultures. She wanted to replicate her awareness of African American culture but also acknowledged she had a long way to go.

> To me, it became knowing the early history of our kids and the different cultures, how they were taught, who was respected in their community, how that has evolved, I think it is a work in progress. For me, it is a very early work in progress. (Postinterview, 6/5/07)

Social justice: A new way to think about mathematics. Inga kept abreast of sociopolitical issues but wrote that she did "not do enough to bring [her] global knowledge to the classroom" (Reflection, 5/7/07). Teaching mathematics and social justice was a new idea for Inga, and she was eager to learn about it. In reflecting on one article, she admitted her unfamiliarity with, but enthusiasm for, teaching social justice and mathematics: "Upon reading the simple way a problem could be worded to make global issues come to light made me see the light. Taking this into consideration is simple yet could be so enriching to my kids" (Reflection, 5/7/07). The readings for the following class affirmed not only Inga's conceptions about teaching mathematics and social justice but about her own practices. "This article made me say to myself that this is more doable than I have thought. It is going to make an impact more than I thought" (Reflection, 5/21/07).

During seminar discussions of social justice lessons from *Rethinking Mathematics* (Gutstein & Peterson, 2005) Inga shared enthusiasm for trying all of the lessons but later expressed caution with regard to bringing sociopolitical issues into the classroom. She wanted to "think of something that would be culturally relevant to our school and our population but I would have to be really careful where it could lead" (Teacher Interview, 6/4/07).

A fork in the road. In her journey toward understanding the relationship between equity and mathematics, Inga seemingly followed different paths. As relates to academic achievement she held fast and even reaffirmed her views about homogeneously grouping students yet identified new ways to learn about and build on her students' out-of-school practices. In terms of cultural competency, she acknowledged her understanding of African American culture, sought to use that understanding in mathematics, and tried to broaden her knowledge of other cultures. Social justice was an epiphany for her, influencing her conception of equity and mathematics. In her final interview, Inga stated,

> I guess I always thought of math as a neutral area and saved my social justice for social studies. When you hear culturally relevant pedagogy with math you are like, what? Math is two plus two. But when you really think about how you are teaching it or how you are pushing it, there is relevant pedagogy in there. There is. But I didn't see it. I thought you were full of crap. (Postinterview, 6/5/07)

Caroline's Story: A Never-ending Journey

Caroline's personal history set the stage for the strong values she developed around justice and cultural understanding. Growing up in a home that embraced diversity and experiencing the race riots of the 1970s awoke in Caroline a deep awareness of the impact of racism: "When history comes to your back door you either ignore it or take notice and I think that even though I was very young it kind of woke me up to this sense of outrage and injustice" (Whole Group, 2/19/07). Caroline is White, her husband is African American, and her children, who are bi-racial, are viewed by most as African American. This presented her with a different perspective:

> I can still personally escape racism under certain circumstances, but I can never escape it far because that which affects my husband and children affects me, directly and intimately. Ugly comments

and anonymous phones calls eventually pale in comparison to the daily accumulation of pain and weariness that racism can bring. (Autobiography, 2/5/07)

This was Caroline's fourth professional development course that explicitly addressed equity in mathematics. Thus, she brought to the group broader personal and academic experience and understanding of how to view mathematics through an equity lens. Caroline was the least likely to use textbooks and based much of her instruction on long-term activities linking the standards to social and cultural issues. Caroline regularly taught with social justice in mind and was thus one of the most social justice–oriented teachers in the group at the time of the seminar. Caroline was very vocal about her personal and teaching experiences and shared these in a positive way. Her story begins with her view of the power of mathematics. Next, she takes her own cultural competence to the next level. Finally, we learn how she comes to view social justice as overcoming inequities.

Academic achievement: The power of mathematics. Caroline had a deep passion for mathematics and its power. In sharing with a small group during one session, she stated, "My expectations I think for math are more global.... It's a way to express yourself. It's a whole bunch of things" (Whole Group, 2/19/07). Caroline's beliefs about what mathematics is and how it should be taught were congruous with the conceptualization of teaching for understanding that incorporates the cultural and sociopolitical contexts of students' lives. This was evident early on, when in her first reflection she wrote, "[Math] is fundamentally about constructing knowledge to build upon and acquire the depth of knowledge to which the learner aspires, be it as informed consumer, mechanical engineer, grade school mathematics instructor, or mathematics researcher" (Reflection, 2/19/07).

Caroline was a strong advocate for heterogeneous grouping, believing it was the best way for all students to achieve academically. As Caroline described her mathematics classroom she explained how she grouped and regrouped her students. For Caroline, heterogeneous grouping was fundamental to teaching mathematics equitably. In reflecting on a discussion in which Inga said she has come to terms with grouping students on the basis of ability even though the majority of the students in the "high" group are White and those in the "low" group are students of color, Caroline stated,

And I heard some things in our last meeting that...I don't like the way that sounds, these are homogeneous groups and then they will say well all the kids were White but it was really great. It is like, no, it wasn't really great. It wasn't great. You need to look beyond that. (Postinterview, 6/5/07)

To Caroline, the fact that the ability grouping resulted in ethnically homogenous groups was enough to deem the practice inequitable.

Despite strong convictions about grouping, Caroline was willing to question other conceptions about academic achievement. She was a firm believer in teaching for understanding but wanted to make sure it was for the right reasons. During her interview prior to the seminar she said,

I know where I stand but I'm always open to, when I hear someone keep saying, "Teach the algorithm first," I want to know why they say that. And I want to know if there's validity there and I want to know why I need to be open to that and when that's appropriate. Because I just don't want to go down my blind path because it feels good to me. (Preinterview, 2/19/07)

Cultural competence: Taking it to the next level. Caroline's personal and professional experiences in examining culture were such that some of the articles the group read and discussions they had were elementary to her. In referring to an article by Tate (1995), she wrote, "It doesn't add too much to say that white culture is the 'default' position of an idealized middle-class reality. I guess I accept that idea pretty readily so I am looking to the next phase" (Reflection, 3/12/07). To clarify her position, she pointed out the importance of focusing in particular on students who struggle the most. "What is important is finding culturally relevant (centric) practices for those students who are struggling or who seem to lack access to the curriculum and where culture appears to be a potential barrier" (Reflection, 3/12/07).

With a foundation in considering culture in mathematics, Caroline sought to take her understanding to the next level through her practice. She thought about lessons and units that would engage students in activities that would enable them to see their cultures and experiences as relevant and mathematical. In a reflection, she wrote, "I have tried to help my students see themselves as part of a larger culture of math and to open them up to an appreciation of the incredible contributions from every culture to the math we study in school" (Reflection, 3/26/07). One example was a unit on international currency. Here, students had the opportunity to choose both the country and the currency value they wanted to research. They were able to see how culture and mathematics were bound together by exploring what currency reflected about the people and the culture of a country.

Social justice: Overcoming inequities. When it came to teaching mathematics *about* social justice, Caroline had some reservations. Her concerns for lesson choice were less about parental and administrative interference and more about striking a balance between building children's knowledge of the world and not painting a bleak picture of the world in which they lived. Caroline was already familiar with *Rethinking Mathematics* (Gutstein & Peterson, 2005) and the lessons we read about during the seminar. She had been doing a form of the world wealth problem from *Rethinking Mathematics* for two years. Yet, Caroline did not want to overdo lessons *about* social justice. During a seminar discussion about what lessons teachers were willing to try in their classrooms, Caroline stated, "I think that the thing about not presenting our world as a big problem is so important" (Whole Group, 5/7/07).

In addition to her concern about children's welfare based on the content they were exposed to, Caroline was concerned about the way content was shared. Although she acknowledged that students were likely to be aware of their teacher's political perspective, she felt it was important to present both sides, stating, "You cannot use them to pursue our agendas. The trick is how do you really engage the students in their interests and help them learn about working for social justice without using them as your tool" (Whole Group, 5/7/07). Caroline not only incorporated lessons *about* social justice, but taught *with* social justice: "A crucial way in which enlightenment and empowerment are present in the classroom has to do with the learning environment that we create and foster. The social contract that I and my students develop is based on social justice" (Reflection, 4/30/07). In teaching *about* and *with* social justice, Caroline's aim was teaching *for* social justice. Caroline wrote, "I have one overriding goal in my teaching and that is to help kids have more power over their lives" (Reflection, 3/12/07).

A never-ending journey. Caroline came to the seminar with greater understanding and experience with equity and mathematics than some of the other teachers, but also recognized that she had a long way to go and always would. To her, the seminar was about constructing the knowledge to add to the teachers' collective understanding about mathematics. "So this boils down to my beliefs about the place of teaching toward a pedagogy of social justice and what I believe that to encompass...my guiding word is not liberation but empowerment" (Reflection, 5/20/07). Here, Caroline emphasized her beliefs about teaching mathematics *for* social justice and the power teachers can provide for students to learn and use mathematics to understand and influence their world.

Discussion

The three teachers came to the seminar with varying attention to and experience considering academic achievement, cultural competence, and social justice in mathematics. These experiences were the lenses for the teachers as they examined their conceptions and practice. Rose's starting point was academic achievement as expressed in teaching mathematics for understanding. From there she broadened her conception to consider students' culture and sociopolitical issues. Inga's experience in a school focused on culture prompted her to identify cultural competence as her strength, even though she had not previously considered how it might relate to mathematics. Caroline started from a perspective of social justice as a result of lifelong encounters with and against racism. As such, she looked at achievement and cultural considerations as integral to achieving social justice. Not surprisingly, the different initial understandings contributed to differences in the teachers' trajectories.

Inga and Rose expressed a newfound understanding of the significance and benefits of considering mathematics instruction through a lens of equity. They both changed their views regarding the importance of incorporating equity into mathematics teaching. Prior to the seminar Inga believed that mathematics was culture free and Rose felt that teaching for understanding was enough. By the end of the seminar their beliefs had changed. Rose came to see how considering both students' culture and social justice could support achievement. Inga's grounding in cultural competence opened the door to build on students' strengths. Although Caroline had already considered equity a critical consideration in mathematics teaching, she continued to refine her understanding during the course.

Each of the teachers identified different challenges with regard to the constructs of culturally relevant pedagogy. For academic achievement, both Caroline and Rose repeatedly expressed the importance of heterogenous grouping for students. Alternatively, Inga felt that "ability grouping" was an effective way to support students' achievement. Rose and Caroline both felt the greatest challenge in understanding students' cultures was the time it required. Although some teachers in the course shopped and attended church in their students' neighborhood as one small part of knowing their students and community, others, such as Rose and Caroline, felt that was tipping the balance between their personal lives and their work. Inga's challenge was language. She had for many years been intimately involved in community activities with her African American students but felt that language was a barrier in better knowing her Latino/a and Hmong families.

For all three teachers, teaching mathematics *about* social justice raised concerns about whose politics were being taught. Each had different concerns about the content used in social justice lessons. Rose raised the prospect of retribution from parents and administration in teaching anything related to the Iraq War. Inga expressed caution about where certain conversations might lead. Caroline was concerned about students' welfare from two fronts. She did not want students exposed to bleak pictures of their world on a regular basis and felt students should not in any way be used a tool to promote or further any teacher's political agenda.

Implications

The purpose of these case studies was to provide the stories of three teachers and their evolving conceptions of equitable mathematics pedagogy. The stories represent the teachers' struggles as they examined their own strengths and areas for growth. Although the results are specific to these particular teachers, they do provide an understanding of different paths teachers followed as they developed understandings of equitable mathematics pedagogy. Using culturally relevant pedagogy as a lens on their mathematics teaching afforded teachers the opportunity to evaluate their beliefs and practices with regard to mathematics instruction.

The overriding finding from the three case studies is that each teacher used her strengths as a starting point or grounding for making changes. Rose's strength in academic achievement was her lens for exploring cultural competence and social justice. She always went back to academic achievement to see how the other constructs could be woven in. Inga's strength was cultural competence. It was from a cultural lens that Inga supported students' academic achievement and evaluated social justice as she linked social justice ideas to students' cultural experiences. For Caroline academic achievement and culture always led back to social justice. Her deep-seated beliefs about the purpose of teaching and role of the teachers as empowering students drove her work. The paths the three teachers followed provide a story about the progress teachers made in this professional development and offer a guide for others as they endeavor to understand and engage in equitable mathematics pedagogy. These teachers demonstrated that regardless of the entry point, there are multiple ways to consider students' cultural and sociopolitical entailments as integral to teaching mathematics for understanding.

As framed, equitable mathematics pedagogy might appear daunting as it may interrupt traditional notions of teaching mathematics for understanding. Much like other research on critical pedagogies (Ladson-Billings, 1994a; Gutstein, 2006) the case studies provide insight into the struggles and issues that others might face in endeavoring to develop equitable mathematics pedagogy. The articulation of the ideas in the case studies may help teachers examine the strengths and areas for growth in their own practice. I anticipate that these cases can be used in professional development and in mathematics methods courses as teachers and teacher educators explore their own understanding of equitable mathematics pedagogy.

In looking across the cases, the analysis revealed an instantiation of the ways in which the teachers' histories and contexts had an impact on how they engaged with the ideas of equitable mathematics pedagogy. I found that (a) experiences and histories influenced starting points and motivations, (b) the teachers portrayed varying degrees of change in conceptions and practice based, in part, on their starting points, and (c) each teacher identified different challenges in engaging in each of the three constructs of culturally relevant pedagogy. From a research perspective, the categories set forth in Table 5.1 provide a frame for researchers to use as they examine the link between teachers' experiences and their perspectives on equity and mathematics.

Reflections on Research in Equity and Mathematics

In this final section I offer my own perspective on the struggles and successes I have experienced in this and other professional development seminars that endeavor to examine equity and mathematics together. First, I feel the need to place myself on Table 5.1. I found this is no easy task, and it confirmed that categorizing individuals in this way does not reflect the whole person. When I began teaching fifth grade in 1999, I had not the embedded cultural or direct experiences with racism that the teachers in groups B and C had. I was, however, aware from an early age of my own privilege as a member of the White middle class and took deep offense to discrimination I witnessed. In that first year of teaching I did not consider any links to equity and mathematics, but it did not take me long to make the connection. Working in a school that started tracking in second grade, I soon saw the inequitable access to mathematics experienced by underrepresented minorities. Thus, I moved quickly through the categories described in Table 5.1. This move provides me with a particular perspective when working with teachers as they endeavor to move in their understanding of equitable mathematics pedagogy. I see where I have been and where we can go.

In terms of professional development, Inga captured the frustration in this work when she assumed I was "full of crap" in looking at a culturally relevant approach to teaching mathematics. Many assume mathematics is culture free and thus we must debunk that myth before exploring how to change our teaching. It is through the persistence of teachers such as the ones I worked with that we can make that change.

Note

1. The names of teachers and students have been changed to protect their identities.

CHAPTER 6

Math Is More Than Numbers:
Forging Connections among Equity, Teacher Participation, and Professional Development

CAROLEE KOEHN

The most promising forms of professional development engage teachers in the pursuit of genuine questions, problems, and curiosities, over time, in ways that leave a mark on perspectives, policy, and practice. They communicate a view of teachers not only as classroom experts, but also of productive and responsible members of a broader professional community.

(Little, 1993, p. 6)

Mathematics achievement is crucial for future national and individual success. While professional development settings cannot bear the entire responsibility for increasing opportunity and raising student achievement, these settings can be avenues for changes in practice. Professional development can provide teachers with opportunities for active learning that equip them to collectively deepen their understandings and take more assertive roles in the education of their students.

Real efforts to effectively educate students will never be fruitful if issues of equity, power, and institutional prejudices are not confronted. Therefore, issues of equity and inequity must be central to efforts to challenge existing discrepancies in student opportunity and achievement (Weissglass, 1999). Martin (2003) posits,

> Equity discussions and equity-related efforts in mathematics education should extend beyond a myopic focus on modifying curricula, classroom environments and school cultures absent any consideration of the social and structural realities faced by marginalized students outside of school and the ways that mathematical opportunities are situated in those larger realities. (p. 7)

Equity is not synonymous with equality. Attention to equity necessitates considering specifics of the context of teaching practice: the environment, students, and student needs. To accomplish this we must understand the larger context of schooling and societal structures (Ladson-Billings, 1998). Furthermore, mathematics is not a culture-free endeavor. Rather, our lived experiences have a profound impact on how we learn, interact, and understand the world (Ladson-Billings, 1997).

Theoretical Perspectives

I set out to create and study a setting that promoted learning through collaborative and authentic experiences, in which professional development was conceptualized as a collective discourse that sought to cultivate a space for practitioners to build professional identity to support practice. *Math is More Than Numbers,* a professional development setting for secondary school mathematics teachers, focusing explicitly on the integration of mathematics and equity, was designed to engage teachers in ways that prompted them to pay attention to issues of participation and access, in order to increase opportunity in mathematics. The following interrelated perspectives guided this approach: (a) situated learning perspectives, (b) opportunities for active learning, (c) development of a critical consciousness, and (d) Critical Race Theory.

Using a sociocultural view, learning and the construction of knowledge are situated within the environment in which one interacts (Vygotsky, 1978). Learners are not passive recipients. Rather, learning is a process of internalization through exchanges with others (Tharp & Gallimore, 1991). People's positional and situated identity shapes their choices, influences their work, and impacts how they make meaning of that work (Martin & Van Gunten, 2002). Varying the context of professional development opportunities yields different possibilities for interaction, engagement, and learning. Providing teachers with opportunities for active learning equips them to collectively deepen their understandings (Little, 1993).

Opportunities for dialogue can promote different outcomes for teachers and students, as discourse practices are integrally connected with identities (Gee, 1989). When teachers are given time to reflect on issues of equity and to collaborate with others, they may begin to view their responsibility as educators differently. Such professional development opportunities provide time for teachers to analyze, reflect on, and challenge their own beliefs and dominant worldviews. This can result in a supportive group of teachers who are committed to dismantling traditional barriers to student success.

Critical Race Theory plays a central role in equity-based mathematics professional development. Critical race analysis in education presents us with different lenses to examine traditional narrow paradigms that exclude students and teachers from enhanced educational opportunities. We must develop ways to acknowledge and value voices and experiences in efforts to analyze "reality" from multiple perspectives and challenge dominant ideologies (Solorzano, 1998). Components of the professional development setting reported on in this chapter engaged teachers in tasks that required them to analyze their current dispositions and share their experiences. Critical Race Theory offers us the mechanism of counternarratives that allows teachers opportunities to create alternative views, explanations, notions, and ways of seeing. These narratives provided teachers with opportunities to tell their stories and create new ways of seeing and a renewed sense of agency (Wiedeman, 2002). I argue that these types of experiences provide a foundation for enacting change in teaching practice.

Goals and Perspectives of the Professional Development

Math is More Than Numbers engaged teachers in issues related to mathematics teaching in urban settings with an explicit focus on equity and access. Specific goals in the design phase helped to center equity within mathematics professional development. The following goals framed the research and supported the ongoing work with teachers in this professional development setting:

1. To engage in critical dialogue around issues of equity, inequity, access, and diversity in mathematics education in urban settings, where attention to equity necessitates considering specifics of

the context of teaching practice as well as the larger context of schooling and societal structures. Discussions of equity will be embedded within mathematics content, pedagogy, and practice.
2. To engage with mathematics problems that incorporate social justice themes and reflect upon our own and student participation in both these and more traditional settings.
3. To adopt a more comprehensive and holistic view of mathematics learning, that includes both procedural fluency and conceptual understanding, as well as strategic competence, adaptive reasoning, and productive disposition.
4. To deepen mathematics content knowledge and pedagogy through a critical lens.
5. To learn about culturally relevant teaching practices as they relate directly to mathematics learning.
6. To examine and analyze teacher practice.
7. To develop an action plan for professional growth.

The creation and facilitation of *Math is More Than Numbers* relied upon a multitiered agenda. The primary aim of the professional development was to support the learning in the room, encouraging participants to draw on their own experiences as mathematics learners and as teachers of mathematics. Second, follow-up meetings with the same group of teachers throughout the school year provided space for supporting continued learning, where participants engaged in social justice–oriented mathematics tasks and ongoing dialogue around equity issues in mathematics education. Third, action plans and other assignments provided a structure for participants to apply this learning to teaching practice and working with students.

This agenda required extensive planning, structure, and flexibility to provide a setting for adults to engage in reflective inquiry. Creating a safe setting for honest dialogue lies at the foundation of this work. It is implausible to expect participants to authentically discuss issues of equity and their own experiences and classroom practice in the absence of a trusting and supportive environment. The absence of trust often results in a threatening environment that prevents the brain from fully functioning. People need to be heard, validated, and supported in their learning and sharing of personal experiences. Professional development facilitators must provide spaces for collaboration that foster empathy, reaffirm listening, and aim to develop understanding, rather than present judgmental stances. These spaces can serve to create the safety needed to empower teachers to be self-directed and challenge traditional mind-sets since dialogue opens minds to other ways of seeing (Costa & Garmston, 2002).

Conceptualizing Equity within Mathematics Education

Equity, as it pertains to the teaching and learning of mathematics, can be examined from at least two perspectives: the mathematics task itself, and the learning environment of the classroom. First, issues of equity were explored while engaging in mathematics tasks. The lessons presented prompted participants to use mathematics to investigate social issues. Traditional mathematics instruction is often fragmented and disconnected. These lessons provided learners with concrete experiences that helped them to realize that mathematics can be an essential analytical tool. Mathematics understanding can be a vehicle for developing a sense of power in transforming society to being more democratic. This opens the possibility for learners to make connections with and develop greater motivation in mathematics (Gutstein & Peterson, 2005).

Second, organization of the learning environment created opportunities for engagement. Instead of relying upon top-down approaches to teaching, *Math is More Than Numbers* draws upon a Freirian approach to education. Learners (in this case, the teachers participating in the professional development) are not blank slates; rather, they must engage in learning. Democratic education negates the idea of a banking system of education, as it stifles authentic learning. Furthermore, educators do not merely teach subject matter. Instead, through their practice, teachers teach how to think critically (Freire, 1983). This professional development provided teachers with active learning opportunities.

Context of the Study

Math is More Than Numbers was a week-long professional development that included four Saturday follow-up sessions throughout the subsequent school year. Morning sessions provided time for participants to engage in activities and discussions about equity, special needs, participation, culturally responsive teaching in mathematics, and building community. Afternoon sessions engaged participants in context-based, social justice–oriented mathematics lessons. Teachers were to do the tasks as students, and reflect on those tasks through the lens of educators. Additionally, as part of her participation in the institute, each person was to identify an area of concern in her own school culture and create an action plan to address and improve that need.

Methods & Data Analysis

A flyer was circulated through a large urban school district to recruit interested teachers. These teachers represented a range of teaching experience (preservice teachers through 30-year veterans), geographical location of teaching (representation from many areas of the greater Southern California region), and racial/ethnic backgrounds. All 14 teachers who participated in the professional development agreed to be study participants. Seven of the teachers were female, and seven were male. The teachers were divided into four groups. Two of the four groups were selected randomly for the purpose of examining group interactions. These groups were the focus of the data collection for the duration of the professional development. In this chapter, I will examine the data from these seven participants.

I analyzed teacher interactions during nonroutine, social justice–oriented mathematical tasks because engagement in these mathematical tasks served as the piece of the professional development that could assist me in understanding teacher learning over time. This study generated data from critical dialogue among teachers around issues of equity, access, and diversity and documented shifts in thinking and participation around equity.

I used video to capture this interaction and critical dialogue among teachers and document shifts in thinking and participation. A stationary camera was situated in the back corner of the room, positioned to view the members of each group during the mathematical tasks. Flat microphones were taped to the center of the table in each group. During analysis, I coded teachers' approaches to the mathematics tasks, areas of struggle, interactions, expression of ideas, and personal reflections. I looked for trends and patterns across the group as well as within individual participants.

To understand how teachers participate, I began by analyzing video of teacher participation within group interactions to look for both participation patterns and substance of participation in two domains: (a) in how teachers talked about themselves in relation to teaching, teaching mathematics, and teaching diverse groups in urban settings, and (b) how they spoke about their own participation as they engaged in the professional development activities.

I was interested in documenting both the ways teachers participate and the shifts in participation over time or activities. I looked at individuals, small group interactions, and whole group settings. This process helped me to understand trajectories of participation over the course of the professional development. This is important, because these participation styles and patterns are indicators of what and how teachers are learning within the professional development. This learning would likely affect their classrooms and the resulting opportunities presented to these teachers' students.

Findings and Discussion

Task Engagement

I analyzed approaches and strategies used to engage in mathematical tasks, from both a group and an individual perspective. Participants worked on the mathematical tasks both individually and collaboratively, although tasks had been designed for group interaction. Participants took more time to both begin problems and engage in collective discourse with other group members earlier in the professional development setting than they did later in the week. In both groups, collaborative work that involved discussing mathematics occurred inconsistently on days one and two, during which more time was devoted to individual endeavors; more group interaction was visible during the middle and end of the activity.

While groups started working on problems together relatively quickly, the nature of the collaboration remained at superficial levels, in that teachers did not readily share their mathematical thinking. For example, on day two, a group of four individuals each came up with his/her own mathematical hypothesis before arriving at a group consensus. Rather than discuss the mathematical thinking behind each hypothesis, the group chose to pick various group members' numbers to report out to the whole group and use as the basis for the rest of the investigation in the task.

Instances of collaboration increased and were enhanced during tasks that required higher levels of cognitive demand. As the week progressed, participants gained more exposure and experience in working through contextualized mathematics problems. Some participants used individual think time strategically to assist them in engaging in group discussion in more productive ways. These individuals needed time to formulate their own ideas before discussing the task with their group members.

As the days progressed, groups worked much more collaboratively to accomplish a shared task. For example, an investigation of mortgage rates on day four was considered by many of the participants as one of the most challenging mathematical tasks of the week. The goal of the problem was to read an article and extract relevant information in order to determine whether or not discrimination was a factor in mortgage loans. While all participants exhibited signs of struggle due to a confusing article, complicated mathematics, and undefined terms, group members immediately began to talk together about what the problem was asking and what information was important to use, while citing instances from their own experiences to help them understand the task. This helped them in collectively solving a task and drawing conclusions that might have been difficult to work out alone.

There appeared to be two reasons for this finding. First, the nature of the problem mattered. Different types of problems required diverse efforts, both across and within individuals. Second, where the task fell within the sequence of days also impacted participation. This finding makes sense, given that none of the teachers in the respective groups knew each other prior to the start of the professional development setting. As participants had increased opportunities to work together and develop relationships, each was more open to engaging in collective endeavors. As the days progressed, participants were provided with opportunities to get to know each other and build trusting relationships within both the small group setting as well as the whole class environment. This atmosphere of collegial trust supported an environment of risk taking for participants to explore more challenging mathematical tasks.

Struggle

In this study, I define struggle as those moments in which individuals and small groups try to make sense of the problem, situation, or given context in their efforts to engage in the task and formulate conclusions. I documented areas in which participants appeared to struggle with either the mathematics or the context of their engagement in the task. Individuals exhibited moments of struggle in a variety of ways. Modes of struggle included, but were not limited to, shifts in body language and positioning, changes in voice tone, extended periods of time of detachment from group discussion, periods of nonphysical engagement, efforts to request individual think time, and verbal accounts.

The mathematics tasks teachers engaged in were highly contextualized and open ended in nature. These types of problems did not suggest a simple or direct method for finding solutions, and often could not be solved with ease. This resulted in various signs of frustration. Analysis of the data indicated two forms of struggle in which the teachers engage: struggle that impeded an individual's progress and struggle that motivated one to continue in her development. This major distinction had a profound impact on how the teachers in this study participated in social justice–oriented mathematical tasks and in their small groups.

Struggle leads to increased openings to participate: The example of Marina. Sharing mathematical ideas and talking through their struggle provided openings for individuals to engage in and understand the mathematics and context of the activity, as evidenced by their verbal involvement and time working on the task. For example, Marina[1] internalized her struggles to such an extent that she engaged in unconstructive talk, verbalizing her negative self-images to her group members at each point she struggled with the mathematics of the task. On day three, she did this 16 times during the task. Her struggle prevented her from fully participating in individual or collective efforts to complete the task and impeded her progress. As a result, she could not formulate a conclusion using mathematics as evidence to support her claims, perhaps because she had not developed an understanding of the mathematics of the task.

During subsequent tasks, Marina experienced heightened comfort and confidence in her participation, and as a result, the number of instances of negative self-talk decreased dramatically. During a task later in the week, this occurred on a single occasion. Furthermore, in this instance when negative self-talk reemerged along with struggle, Marina was able to shift her perception of being unable toward productive work using mathematics. She utilized problem-solving techniques to make sense of the context and mathematics and explored various methods for arriving at a mathematical solution. She did this by verbally working through the problem, in collaboration with others. In the whole group setting, Marina reflected on her participation across days in the institute,

> I had a different experience today. For some reason, yesterday's activity and today's activity were both of high interest to me, but the mathematics and what we're being asked to do was more

difficult for me yesterday than it was today, like there was enough information in this article that, like for example, when we had to explain what the disparity ratio was, I had all the numbers and I could play with them so that I would come up with that ratio, whereas yesterday, we had to set up the simulations and we had to construct something from not really having anything right there given to us, so it was a lot more challenging. Yet, I felt like I was able to participate more today and I was a lot more of a participant today because of, of my comfort level with the activity.

Marina was clearly competent in her mathematical abilities. However, moments of struggle impeded her progress early in the institute. Later, she experienced a shift in motivation and determination in moments of struggle.

Struggle as mode to engage. None of the participants in this study, in times of struggle, continually remained in a place of motivational struggle. Rather, all experienced transitions in their participation structures. For example, each participant experienced moments of struggle during at least one mathematical task in the professional development. However, as they engaged in more tasks and as the institute progressed, teachers gradually began using their own experiences to generate meaning, make sense of the purpose of the activity, and engage more fully in group discussions. Struggle was not always negatively conceived or directed in unconstructive ways.

In each group, though to different extents, a group member vocalized her struggle and lack of understanding to other peers early in the problem-solving process, after she had spent some time thinking about the problem independently. The group was then able to discuss the context of the problem, generate examples from their lived experience to understand both the context and the mathematics, and brainstorm ideas for solving the task at hand. Under this scenario, persons were no longer working as isolated individuals. Rather, a high level of group functioning existed, and was essential to each person gaining a clearer understanding of the task. This is in contrast to superficial ways in which groups sometimes operate, where there is a distinct, continual leader and each member is not a fully participating or valued member of the group.

Over time, participants appeared to benefit from collaboration with others in developing a repertoire of problem-solving strategies. They worked together on a shared task, drawing upon each other as valuable resources and developing identities as doers of mathematics through shared practice (Wenger, 1999). Collective work provided a space for teachers to share their mathematical thinking and approaches. Talking through their struggle provided openings for individuals to engage in and understand the mathematics and context of the activity, as evidenced by their verbal involvement and time working on the task. Through collaborative efforts, more participants persevered in working through tasks even when struggling with either the mathematics or the context.

Struggle: Creating potential spaces for redefining identity and positionality. The tasks in this professional development setting were purposely designed to engage participants with contexts that challenged traditional notions of mathematics and mathematical tasks. The data show that groups and individuals within groups were better able to understand the task and draw conclusions when group functioning was most collaborative. Thus, exploring a diversity of viewpoints and experiences promoted deeper understanding of both the mathematics and its value in analyzing social issues.

The notion that each member played a vital role in contributing to the collective understanding of mathematical tasks created a different expectation for group dynamics and collaboration. Groups no longer functioned as tiered individuals, with one clear leader and the other group members as followers. Rather, more participants began to see themselves as doers of mathematics. This was important in enhancing both individual and group understanding in the mathematical tasks. Group dynamics and previous experiences can greatly influence individual participation. In this setting, teachers participated more readily and took more mathematical risks after they had opportunities to develop relationships with their group members. This has profound implications for participant identity and positionality both in the group and within mathematics.

By the end of the week, each participant was able to solve mathematical tasks and draw conclusions that connected the mathematics with the implications of the social issue being investigated. Learning mathematics occurred as a result of participating within a social context. The nature of these open-ended mathematical tasks required learners to consult with others in ways that promoted deeper understanding of the mathematics and the context of the task.

I argue that the process of engagement was much more important than arriving at a correct answer. All teachers in this study evolved in their participation structures and grew in their ability to work on social justice–oriented mathematics problems. This resulted in most participants developing a stronger sense of mathematical identity that allowed them to see themselves as a contributors to both individual and collective mathematical understanding. By the end of the week, individuals did not blindly accept others' mathematical ideas. Rather, they offered conjectures of their own. This had implications for participants taking ownership of their own learning and possibly built confidence in their mathematical ability.

Transformative Shifts

Each participant in this study faced moments of discomfort, insecurity, or struggle due to mathematical tasks that necessitated complex thinking, yet required no prescribed way to participate. To varying degrees, each participant experienced a transformative shift as a result of her growth in this professional development setting. Here, I define transformative shift as a change in composition, structure, character, or condition in one's way of seeing that prompts one to assume responsibility. For some participants, this happened during the initial professional development. For others, the transformative shift occurred later in the school year during the follow-up professional development sessions and/or in the action plan implementation while working with students.

Once a person experiences a transformative shift, she creates a new way of seeing and does not revert to previously held long-standing notions. One participant shared this during a follow-up session,

> I am beginning to see that my so-called advanced students aren't any smarter; they are just better at doing school. I now see my role as a teacher to create opportunities for all my students to learn math and share their understanding.

As a result, this teacher began to change the learning environment in his classroom to be more inclusive and collaborative. Lessons that previously focused on lecture and practice with corresponding notions that students needed to conform to teacher expectations, transitioned to investigations that prompted students to use mathematics in context, develop conceptual understanding, and apply problem-solving strategies.

Participants' Reflection

I documented (a) instances in which participants made personal reflections on the activity or on the manner in which they participated in the task, (b) moments when individuals or groups used their own experience to generate meaning and reflect on learning, and (c) connections participants made to previous experiences in professional development. I was especially interested in the timing of reflective comments during each mathematical activity.

For all seven participants, reflective comments, especially those that utilized personal, lived experiences, occurred at the end (if at all) of the activity during the first days of the professional development. A transition took place during days three and four: teachers identified personal experiences at the beginning and throughout the mathematics activity. This allowed the teachers to use these reflections to help make sense of, conceptualize, solve, and draw conclusions using mathematics within the context of the task. This was of particular importance considering that all of the teachers reported that they had no previous experiences engaging in mathematical problems of this type, where the mathematics investigated originated from a societal context, as opposed to starting with the mathematics and developing a word problem to simulate a real-world example.

The number of reflective comments per individual varied substantially, although five of the participants were fairly consistent in the total number of reflective comments across days. Because reflective comments are made when a person actively verbalizes what is formulated in her mind as a result of her participation, this indicates that teachers were in different places in terms of being able to connect the mathematics they engaged in with the context or possible implications of the activity.

Teachers' participation evolved over the course of the institute, as they engaged in more reflective thinking. Most teachers, although to varying degrees, were able to reflect on the relevance of problem posing and using their own experiences to help them connect to how their students might participate, respond, do, and engage in these as well as more traditional types of mathematics problems. This indicates that these teachers were beginning to make connections to teaching practice.

A majority of participants in this study were able to reflect on their experiences as learners and as teachers. To varying levels, they used their experiences to make connections to students and classroom practice. Participants who engaged in reflective practices were able to make connections between their learning and student learning experiences, as evidenced by their verbal comments. The nature of teacher participation evolved as teachers became more reflective and utilized reflective strategies in ways that helped them understand and complete the task.

Implications for Professional Development

Creating spaces where all participants can meaningfully engage in collective inquiry and group functioning is a vital component of equity-based mathematics professional development. Therefore, it is important for professional development facilitators to establish relationships with participants that allow the facilitator to be a part of the learning community. Facilitators must design meaningful professional development opportunities in a supportive learning environment in ways that support the development of positive mathematical identities and encourage participants to draw upon their complex experiences to assist them in understanding mathematical ideas.

In this endeavor, facilitators should expect a variety of forms of engagement and problem-solving approaches, predicting that participants may respond differently than expected to the content or task. Facilitators must be flexible in meeting the needs of the group, attempt to identify beforehand instances in which struggle may surface with an intent to become skillful at managing and assisting in those times, and learn to recognize the various forms and patterns of struggle that might emerge. These foci can assist facilitators in developing a better understanding of teacher participation. While it may appear counterintuitive, allowing participants to struggle can lead them to increased understanding and a heightened sense of mathematical facility.

Teachers experience a multitude of pressures in their professional lives. As a result, external messages are often conveyed about what they have to do. Yet, teachers are rarely given the tools or time to effectively address those particular issues. However, individuals' personal learning experiences are connected to their participation in work within mathematics and the resulting discussions with their colleagues. Therefore, teachers need opportunities to reflect and make their own connections. Personal connections are more meaningful and have a greater likelihood for more substantial impact on practice and student opportunities in school and classroom environments. Structured space for written and verbal reflection affords participants opportunities to draw upon their experiences as learners and as teachers to envision experiences they hope to create for students.

Nontraditional mathematics tasks have the potential to change ways in which people participate. These opportunities created spaces for teachers to analyze their own participation, which allowed them to pay attention to their own notions of opportunities and participation.

Consequently, teachers were then able to make connections to their students and pedagogical practice.

Reflective opportunities benefited participants in at least two distinct ways. First, reflection helped teachers to make sense of the mathematical task they were asked to engage in. Second, reflection provided a forum for teachers to use their experiences—both previous experiences as well as experiences in this professional development setting—to make connections to their own learning as well as the learning of their students. I believe a central focus of equity-based professional development must be the creation of safe spaces to encourage teachers to engage in reflective inquiry that includes exploring self in relation to teaching and the personal impact each has on the learning environment he/she creates.

This form of equity-based professional development cannot dictate to teachers that a transformative shift will occur. Rather, these shifts occur naturally over time, within a structured and supportive environment. Moreover, it is impossible for anyone other than each individual participant to decide where he will focus his attention and what areas of his teaching practice will be affected. These shifts depend on multiple factors, such as current dispositions, previous experiences, and the beliefs that individuals possess as well as their openness to listening to alternative viewpoints and engaging in reflective dialogue. Equally important is the interaction among participants and between participants and facilitators, as the nature of participation with others can have a significant impact on one's trajectory.

Conclusion

Math is More Than Numbers was not a make-and-take workshop. Rather, its equity-minded approach required deep reflection. As teachers participated in this equity-oriented professional development, they had opportunities to negotiate their mathematical understandings, reflect upon their pedagogical decision making, and focus on those areas they would like to positively impact. The enhanced sense of agency in this setting supported the creation of different orientations and actions in teachers.

These results cannot be achieved in one-shot workshops or narrowly defined professional development offerings that focus on isolated skills or strategies. Teacher dispositions, entrenched beliefs, and structural inequities are not transformed without difficult, multifaceted dialogue and time for purposeful and conscious rumination. Therefore, professional development must be long term in nature and provide teachers with opportunities to critically analyze existing conditions and possible alternatives. This necessitates spaces for collective inquiry and honest reflection.

My conceptions of equity-based mathematics professional development encompass at least two foci. First, content that includes contextualized, meaningful mathematics tasks and activities that engage participants in dialogue around equity issues centrally related to mathematics teaching in urban settings. Second, equity-minded pedagogy is crucial to the process of engaging participants in such dialogue and reflection. Pedagogy refers to the structures and techniques of approaching the facilitation of professional development with a vision of utilizing diverse measures to increase opportunities for underrepresented groups in ways that value and respect diverse voices. This requires a reconceptualization of professional development practices to position teachers at the center of the work in ways that would allow them to draw upon their collective expertise and aim to meet the diverse needs of their students. These endeavors necessitate a structured way to engage teachers in critical reflection and transformative action around issues that relate to their setting and students.

Professional development and spaces for collective inquiry have the potential to provide educators with openings to challenge existing inequities. Teaching and learning mathematics is not a culture-free endeavor. Realizing that one's experiences shape how one sees, understands, and participates in the world can help a teacher to develop a critical consciousness in her approaches to engaging her students (Ladson-Billings, 1997).

While this was not a causal study, I contend that these teachers' participation in *Math is More Than Numbers* helped them to develop a sense of responsibility and the capacity to increase opportunities for traditionally underserved students. Teacher agency, or the belief that one can impact change, can be seen through the actions of these educators. Many of the participants now believe that what they do as a teacher affects the classroom environment and potential outcomes, and are taking actions to effect those transformations.

I have found that how teachers take up issues of equity and access is highly personalized to their specific needs and contexts. Teachers are more likely to work on small changes in their daily practice where they feel they can have an impact or in areas that are of particular interest to them. In my work as a teacher and professional development facilitator,

I have noted that opportunities for this type of professional development are rarely available in schools. The challenge I hope to raise in this research is to invite others to envision different alternatives to the traditional paradigms they face that inhibit meeting the needs of the students they serve.

Note

1. This name is a pseudonym used to protect the participant's identity.

CHAPTER 7

Using Artifacts to Engage Teachers in Equity-based Professional Development: The Journey of One Teacher

KRISTINE M. HO

Over the past few years the educational community has shifted its focus toward understanding and evaluating the educational experiences of students of color. In particular, there is a cadre of mathematics researchers who have begun to interrogate notions of equity and the impact of race. In the process of investigating race's impact on equity in the mathematics classroom, many questions have surfaced, particularly in the area of teacher development. Engaging teachers in professional development (PD) that addresses the impact of race on equity in the mathematics classroom is a complicated and challenging task (see Battey, Foote, Spencer, Taylor, & Wager, 2007; Foote, Bartell, & Wager, 2007; Franke & Kazemi, 2001; Gutiérrez, 2002; Kazemi & Franke, 2004; and Wager, 2008).

This chapter details the journey of one teacher during a year-long PD effort in Los Angeles. The focus of the PD was to engage teachers around various artifacts to address and discuss issues of race and equity in their mathematics classrooms. In the beginning of this chapter, I provide an overview of current literature on race-centered mathematics PD as well as a theoretical framework that helps make sense of the journey of one particular teacher. I then detail the shifts and changes this teacher made as she engaged with different artifacts. I conclude with a discussion of how these shifts inform the field on race-centered mathematics PD.

Current Research

Race-Centered Mathematics Professional Development

PD has been developed on the basis of various theories of how teachers learn. Sociocultural theory describes interactions between individuals in a community of practice; this provides a means to investigate how the relationships, norms, and identities of and between individuals are significant to foster teacher learning (Lave, 1996). To effectively support and prepare educators of students of color, a genuine microanalysis of race, norms, identity, and context, all central to sociocultural theory, is necessary.

There is a lack of information to aid in the comprehension and development of attention to race/equity within mathematics PD. Teachers and students need a way to discuss and formalize methods that help low-income students of color receive a more equitable education. Teacher learning should help teachers develop ideological and political clarity so that they "see themselves in solidarity with their students in their students' communities" (Bartolomé & Trueba 2000). Furthermore, in order to gain a deeper understanding of experiences in learning mathematics, it is necessary to explore the broader contexts within which race and equity are situated (Martin, 2003).

Race is "one of the most, if not most, salient framing characteristic for differential achievement in the discourse that surrounds the 'achievement gap'" (DiME, 2007). To design PD that effectively addresses this issue, further work is needed. In a study investigating effective ways to help teachers in critical reflection, Howard (2003) says,

> The racial and cultural incongruence between students and teachers may be another factor that explains school failure of students of color. Teacher practice and thought must be re-conceptualized in a manner that recognizes and respects the intricacies of cultural and racial difference. Teachers must construct pedagogical practices in ways that are culturally relevant, racially affirming, and socially meaningful for their students. (p. 197)

Developing these pedagogical practices is critical in the process of identifying and addressing the impact of race on the educational and learning experiences of students of color (Howard, 2003). As a result, if mathematics PD is to interrogate the agenda on equity, race must be at the forefront of such an agenda.

Culturally Relevant Mathematics Professional Development

If we do not actively reform, reconceptualize, and redesign PD so that it focuses on race, equity, and mathematics, low-income students of color will continue to receive an education that both marginalizes and oppresses. Furthermore, passive acceptance of dominant deficit models will not be questioned, challenged, or confronted. The absence of a focus such as how to address culture and race in past PD has led to insufficient change in mathematics pedagogy. In a chapter produced by Diversity in Mathematics Education, Center for Learning and Teaching (DiME, 2007), the authors discuss the urgency of making connections between in- and out-of-school mathematics practices. It is through these connections between schooling and home life that teachers are able to develop curriculum that is more engaging for students (Ladson-Billings, 1997). More specifically, culturally relevant pedagogy is the means through which teachers show students how to "maintain their cultural integrity while succeeding academically" (Ladson-Billings & Tate, 1995). Future PD must inform, prepare, and support teachers of low-income students of color to be reflective about their practices and effectively deal with the effects of race in the mathematics class.

One method to help teachers recognize the need for change is to expose them to the realities of the lived experiences of their students. If teachers are able to have contact with individual students who challenge their underlying deficit ideas, they may be able to question them. The majority of teachers in schools that serve low-income students of color are White (Sleeter, 1997), and these teachers often hold "colorblind" perceptions of their students and rarely support the notion that their low-performing students of color can and should achieve at high levels similar to their suburban White counterparts. Confronting these ideas may be effective in supporting teachers to take steps to develop "social and political" clarity (Bartolomé & Trueba, 2000).

Creating opportunities for teachers to reflect on and discuss student participation, equity, and race may facilitate teacher growth. Creating opportunities for teachers to discuss and unpack their current perceptions of students as racial beings will help them to position themselves to develop culturally relevant classrooms.

Methods

My methodological approach was a design experiment that documented the efficacy of particular artifacts in creating structures for discussing issues of race and equity in ways that shape teacher identity and support a change in teacher practice. Throughout the course of the PD, I investigated how to engage teachers in meaningful ways around issues of race and equity in the mathematics classroom.

In this PD, I engaged a group of teachers in district-based, ongoing PD. I then collected data, and on the basis of that data made changes in the use of particular artifacts to support teacher engagement on issues of equity. There were five focal themes within the PD: (a) community building, (b) challenging deficit notions, (c) defining equity and equitable teaching practices, (d) examining student participation structures, and (e) making sense of teacher practices. Within each theme, I engaged teachers in activities around artifacts that attempted to expose and promote teacher learning.

School Setting and Participants

Schools. The PD took place at Southwest High School and Northwest High School,[1] two public schools located in a small district southwest of Los Angeles. Both schools shared similar demographics and challenges in helping students meet state-mandated benchmarks. The data on which this chapter is based was gathered at Southwest High School. Of the approximate 3,089 students attending Southwest, the majority were Latino, with about a 30 percent African American population, a less than 2 percent Caucasian population, and the remaining students of various Asian ethnicities. Southwest was classified as a Title I school, with about one-third of the students being English Language Learners. Of those, a majority were Spanish speakers (Centinela Valley Union High School District, 2006). Low performance on high stakes testing initiated various programs and proposals to improve student achievement. Unfortunately, many of these programs have not yielded higher test scores.

Participants. To recruit teachers for my PD, I met with the entire mathematics department to present the opportunity of participating in the PD. Five teachers agreed to participate. Each teacher came to the PD with a range of experience and sensitivity toward issues of equity and race. Although the interactions of the group as a whole provided important insights on a teacher learning continuum, one teacher in particular exemplified the common struggles and tensions teachers of low-performing students of color negotiate. This teacher's growth over the course of the PD created many opportunities to analyze and question the efficacy of using specific artifacts in teacher learning.

One distinctive teacher. Sophie was a White American female in her sixth year of teaching mathematics at Southwest. She taught various levels of mathematics from beginning Algebra to Algebra 2. She began teaching directly upon completion of a Masters and credentialing program that focused on issues of social justice. Sophie came from a privileged background that provided her with schooling experiences that were quite different from those of her students.

Context of the Study: Artifacts

Professional development design. Throughout the PD, I engaged in periodic reflective analyses of data collected during each session. Consistent and frequent reflection and analysis provided opportunities for restructuring and action. Three components shaped the PD design: (a) initial preparation and hypothesis articulation, (b) actualizing and experimenting of proposed accommodations, and (c) reflective analysis (Cobb, 2000; Confrey & Lachance, 2000).

The PD included a progression of activities and discussions that I hypothesized would challenge teacher practices and identities while the teachers negotiated race and equity issues. More specifically, the PD supported teachers in extending their notions of what it meant to teach mathematics for social justice by engaging them with various artifacts to help transform their practice. Finally, I tried to guide teachers to think through issues of race and equity in relation to their own classroom practice and the societal structures that shape student opportunities.

Community building. To engage teachers in a PD that closely examined how race emerges in teacher practices, building a sense of community was essential (Grossman, Wineburg, & Woolworth, 2001). The first artifact introduced in the community was an autobiography. Participants were asked to write and share an autobiography describing experiences in which mathematics, race, and equity intersected. Autobiographies of this nature help teachers ground their experiences and make comparisons to student experiences (Connelly & Clandinin, 1990). Teachers then shared their autobiographies with the rest of the group. By exposing different perspectives and providing a space to investigate teacher identity shifts through their positionality, teachers learned more about the lived realities of other teachers, thereby supporting development of the learning community. As teachers began the process of understanding other group members, they recognized some connections between these stories and the stories of the students in their classes. Over the course of the PD, we referred to the autobiographies and the conversations around them. They grounded our discussion about equity in the lived experiences of the teachers while noting how their experiences might be the same and different from those of their students.

Challenging deficit notions. As an additional means to develop community, I engaged teachers with a number of texts that spanned a broad range of issues of mathematics, race, and equity. Video texts as well as articles were used. The texts served to challenge their notions of how they related to their students. During the community-building portion of the PD, teachers were encouraged to create spaces in which they developed authentic relationships with their students. As teachers were faced with a mismatch of ideal and realistic experiences of their students, they reconciled their role in trying to bridge the discrepancies. Furthermore, the space created allowed for a more personal look into students' lived experiences, and therefore, supported teachers in developing a more personal relationship with their students.

Student participation structures. A theme that was explored during each session was the challenging of teachers' notions of student participation. For this, teachers interacted with another artifact, student work, which provided opportunities to investigate student understanding and participation, thus further assisting teachers in the creation of more equitable practices in their classrooms.

During the fourth PD session, teachers were given three pieces of student work. Each piece of student work presented a solution to the same mathematics problem; low, medium, and high levels of understanding were each illustrated through the work samples. Teachers were then asked to look for evidence of student understanding. They were given individual time to make note of what they believed the student understood; they then assigned a low, medium, or high ranking. During this portion of the activity, teachers were encouraged to draw on their experiences and elaborate on their notions of student misconception. They were then given time to share with a partner what they found. This was followed by a group discussion around student work and the ways to assess understanding. After the group discussion, teachers reflected individually in a final reflection responding to questions about the student work analysis.

This was the first opportunity in the PD for teachers to examine student work. The positive feedback from some teachers as well as the deficit comments on student work suggested a need for further analysis of work. The teachers in the group brought in subsequent work samples for extended sessions of student work analysis. Teachers also focused on articulating what evidence they expected to see of student understanding.

Making sense of teacher practices: Curriculum study. The next artifact I introduced was what I called "Curriculum Study." On the basis of the conclusions they made as a group through analysis of student work, teachers developed a lesson around a topic in the unit they were all teaching. Teachers were given time as a group to brainstorm a lesson that would help meet the immediate needs of their students. After they created the lesson, all teachers taught the lesson before the next PD meeting. During the next session, time was spent in small groups and then as a whole group to debrief how the lessons went and what modifications the teachers would make for the future.

Results: Sophie's Journey

At the onset of the PD, Sophie held both socially just ideals for teaching mathematics and deficit views of student learning. Although Sophie was very vocal about wanting to make positive changes for her students, she struggled in recognizing the implications of race and equity. Her desire to be a supportive and reflective teacher was great, but there were several barriers she encountered during our discussions. Some artifacts incorporated during the PD were more successful than others in engaging Sophie around issues of race and equity. The following section will provide more details about these artifacts and Sophie's responses.

Sophie's conflict between socially just ideals and deficit notions were recorded during our first session together. During the second session, she began with a reflection of her schooling experiences and the intersection of race and equity by describing how the lack of diversity in her schooling experiences had shaped her worldview, and how she was trying to resist those ideas:

> I felt growing up, that even though my group of friends are diverse...I have a Jewish friend, a Muslim friend, but we all came from the same background...I was itching to meet someone I hadn't known since kindergarten....I wanted diversity in a school in a college, but honestly when I decided to do the program, I had no clue what social justice was. (Audio, 12/12/07)

As Sophie stated in her mathematical autobiography, she recognized the disparity between her high school experiences and those of her students. She recalled her fear of how students would respond if they knew her privileged upbringing:

> I feel that writing my autobiography brought back a lot of the disconnect I used to feel with my students. I felt ashamed that I had gone to [a school for affluent students] and I didn't feel like I had a place telling them what to do. (Written reflection, 12/12/07)

The autobiography reminded her of the disconnect she had once felt with her students. The differences between the experiences Sophie had and those of her students coupled with her underdeveloped notions of socially just education may have been reasons for the internal struggle Sophie faced.

Internal Conflict

Within the reflection from the second session, Sophie elaborated that her time as a teacher in an urban school gave her a reason to challenge the deficit notions of friends and family (Written reflection, 12/12/07). Sophie's description of how she responded to negative reactions of her friends to her teaching in an "urban" school was one of the initial instances of Sophie battling deficit notions and embracing more socially just ideals. Yet, deficit notions surfaced as well throughout the PD. The following occurred during a discussion around White flight. As teachers began to discuss the impact of a new school being built in the neighborhood that would pull the White and more affluent students from the district, Sophie described her uncertainty of what she would do when she had kids. Deficit notions of the behavior of students of color permeated her commentary. She emphasized how students of color are negative influences and attempted to rationalize this perception by claiming they were not based on race.

> I was also frustrated with the discussion about White flight because I have thought about if I had a kid, I would not send my kid to Southwest HS. I think what I have seen from the kids is a lot of peer pressure. I think the teachers here are probably better than at [higher-performing high schools], but I feel like a lot of it is peer pressure...I wouldn't want my student [at SWHS] either, and that doesn't have anything to do with the race thing. I wouldn't want my kids in a school where students are such bad influences. (Audio, 1/9/08)

At the end of this discussion, Sophie still did not see how her comment reflected deficit and racist notions.

Signs of Alternative Explanations

Sophie offered various explanations for students' lack of success. During a conversation in our second session, Sophie suggested, "The one thing that was missing was a discussion of the relationship of tracking race to socio-economic status" (Written reflection,1/9/08). This is the first instance in which Sophie tried to relate racial issues with socioeconomic ones. Another explanation Sophie suggested attributed low performance among students of color to a lack of value for education: "The thing that shocked me when I started was that even the good kids sometimes failed.... It's not just the dropouts that don't care" (Audio, 12/12/08). In other words, Sophie had a perception of ways students exhibited "caring" about school. She believed that many students did not demonstrate a regard for school and subsequently blamed failure on students' lack of effort. Sophie also talked about her persistence in schooling and how peer pressure to "do well" in school had motivated most of her classmates (Debrief field notes, 1/09/08). She was convinced that persistence was an indicator of success. Since even the students who she deemed as "good" did not persist at all times, that translated as a lack of caring about their schooling. Explicitly discounting any connections to race, Sophie continued to focus on alternate explanations for inequity. Her redirection created obstacles for an authentic analysis of the structural, systematic, race-based factors that cause inequities.

Sophie's perception of parallel arguments for race and socioeconomics were instances of avoidance. As she shifted focus toward using alternative, nonrace-based explanations for inequity, she inhibited her understanding of race's impact. Specifically, Sophie wanted to place more responsibility on the student for lack of motivation and low performance. In addition, she claimed peer pressure would "encourage" higher achievement. Ultimately, Sophie believed that the onus was on the students, and their inability to motivate their peers was problematic.

Thinking about the Importance of Race

In an attempt to support the development of Sophie's thinking about student behavior and its relationship to race, I asked her in front of the group if a student's race might lead someone to interpret behavior in particular ways. Sophie agreed, saying, "Yes, that is possible" (Audio, 1/9/08). This acknowledgment illustrated a slight change in that she was willing to entertain the idea that the race of a student may affect how he/she is treated.

When the group discussed the labeling of students on the basis of performance, Sophie described in her reflection a small shift in how she saw race play out:

> I feel that there is a little bit of a chicken/egg thing going on with race vs. Algebra Essentials classes. It's hard to say if students' race and cultural attitudes are not being addressed properly or if the courses are tracking students and causing a racial divide. I do feel that the label of the class will have some effect on the mental attitude of the students but perhaps making them feel "advanced" or as if they should be in college prep classes will lead them to eventually enrolling in real college prep classes. (Written reflection, 2/6/08)

In this written reflection Sophie recognized the possible interconnectedness of race and performance. These were initial indicators that Sophie was moving toward recognizing the salience of race in equity.

Deficit Notions Revisited

Although Sophie displayed some growth, she continued to voice deficit notions. During a discussion about how labels translate into generalizations about student behavior and learning, Sophie's remarks highlight her essentialization of African American students:

> I don't think those terms are interchangeable...kinesthetic and African American students. There's a reason I'm using kinesthetic; maybe it does happen to describe a lot of my African American students or whatever group I'm trying to describe. It may have some racial similarities...obviously we all have noticed that certain students of different races sometimes tend to act a little bit differently in classes and the culture is different based on that...and sometimes you need to discuss that learning style. And, I don't think I'm using it to mask talking about race. (Audio, 2/6/08)

She began the statement wanting to reject deficit notions, but ended up relating behavior to cultural factors.

Sophie also struggled with defining what it meant to be a "socially just" educator. She had difficulty thinking about what it meant to investigate teaching practices that would create more effective experiences for her students. Although Sophie expressed increased interest in finding ways to create better mathematical experiences for her students, her ideas often included ones that were both deficit oriented and socially just. Reading Ladson-Billing's "It Doesn't Add Up: African American Students' Mathematics" (1997) created an opportunity for Sophie to realize that she engaged in Ladson-Billing's definition of "good" teaching as well as a "pedagogy of poverty." She explained, "I felt like I kind of read through the list. I do that, I kind of do that, if I have time.... Then I went through the pedagogy of poverty, and I kind of do both. I do all the good teacher and bad teacher things" (Audio, 4/2/08). At this point, Sophie understood that there were opposing views, both of which she held.

Recognizing a Relationship between Race and Equity

Reading an article by Rousseau and Tate (2003) encouraged Sophie to recognize a relationship between equity and race. In the conversation about building mathematical identity, Sophie showed a shift in her thinking of how race manifests in inequities for students of color:

> Seventy percent of our students walk into our class being their least favorite from second one, and we have to fight against that. Our math curriculum can be so contextually removed from student experiences, thus, resulting in misunderstanding and at times lack of engagement from students of color. There is a problem in geometry about a silo.... [students] get stuck on it so much on that, it says it's cylindrical, it says what the base is, but they can't move past it and they always skip that problem because they don't know what a silo is. (Audio, 4/2/08)

In this interaction, Sophie saw how students who do not have the same norms and experiences could misinterpret contextual cues. Just as Sophie showed signs of a shift in thinking, however, deficit notions emerged again.

Old Ideas Die Hard

During the tenth session we discussed an article that suggested equity was an outcome, not a process. This raised issues for Sophie because she interpreted this to mean that one race could get "more," and

> if you have two races that aren't scoring them equally, then you need to do different things for them. I'm OK with different, but I felt like if you do more for one race then it meant one race was more superior...Like having an African American club. (Audio, 4/23/08)

Sophie found it unfair and problematic that a club would serve only one race, placing more importance on that race. Later, in the same conversation, Sophie revisited the idea of meritocracy and that her African American students needed to try harder. She brought up an example of looking over test scores and responding to her African American students' outbursts in class:

> I did bring it up to my class, this is what the test scores show, but it doesn't reflect intelligence.... I think a lot of it has to do with some students aren't just trying... whether they are not interested, or the education doesn't reflect it, that's another issue. (Audio, 4/23/08)

Sophie recognized that test scores did not accurately measure intelligence, a more "socially just" notion, but in the same thought placed the blame on students not "trying." Sophie continued to struggle between deficit notions and being "socially just."

Reconstruction of Philosophy of Socially Just Mathematics

Sophie's inability to recognize the impact of race on inequity limited her growth. The small movements she made, however, may have created opportunities for Sophie to redefine "socially just" mathematics. Sophie's biggest shifts were found in her conceptualization of "socially just mathematics." Sophie seemed to wrestle with reasons for inequity in the mathematics class, but toward the end of the PD she expressed how the study had helped her to redefine social justice mathematics. Throughout the process, Sophie was interested in creating/defining social justice mathematics teaching:

> Is it what they're learning is social justice or how they are learning. It made me feel so much better, by maybe doing group work, by

having discussions about it...and still have it about long division of polynomials....Now I can see that...there are going to be different forms of [social justice mathematics]. (Audio, 2/20/08)

Sophie redefined socially just education by focusing on the method rather than the difficulty of mathematics. She also recognized the possibility of different interpretations of "socially just" mathematics. The last two sessions provided more evidence for Sophie's reconstruction of "socially just." In response to an article by Birky, Chazan, & Farlow (2008) that discussed successful teachers of African American students, Sophie's written reflections reiterated her development of different definitions of social justice mathematics:

> I like my new definition of social justice (that students have access to advanced mathematical concepts) better than my old definition (all lessons are culturally relevant and increase student awareness of social/racial/gender equity). While I think that both ideas have to be incorporated together, I think that in my math lessons I lean toward the new idea. (Written reflection, 4/2/08)

She commented on how the discussions and readings in our meetings together had helped shape her new and more attainable ideal for what "good" socially just mathematics teaching looks like. Yet, Sophie felt like she lacked a culturally relevant curriculum:

> I guess the cultural relevance part of it...that is my fear. That it's the socially just part that I'm not doing. Maybe we had a great math lesson; I helped you learn this lesson....I think I was a little nervous about the article because I was concerned that I was one of the "new" teachers that was inappropriately ignoring race....I felt much better. I do think that I offer students of all races similar opportunities, but I believe that we as teachers in our district must offer many more opportunities than those in [more suburban districts]. (Audio. 4/23/08)

Sophie appeared to recognize deficit-oriented aspects of her practice, but immediately justified her actions.

Beginning Shifts in Teaching Practice

Along the process of redefining socially just education, Sophie exhibited inconsistencies in her journey. Readings and discussions were a catalyst to question preexisting practices that yielded small shifts in her practice. Sophie self-reported that she changed the ways she used verbal and written language with her students. She tried to make all encounters more "student friendly":

> I think that I'm a little more careful in what I say or write. Since our discussion on language and our problem analysis, I have been thinking about if I am using math language or if in an attempt to make my language student friendly, I'm actually doing them a disservice. I don't think I have made any major changes to my teaching, but I have thought a lot about what I am teaching and how I approach it. (Written reflection, 3/5/08)

She emphasized that our meetings together resulted most noticeably in how she thought about what she was teaching and how she approached her teaching.

Furthermore, Sophie used various foci we had employed in our meetings to look at student understanding and center discussions in her classroom (Debrief field notes, 4/2/08).

In a group discussion during the last session, Sophie showed additional signs of small shifts in her thinking. She discussed her awareness that people made assumptions on the basis of race, but suggested it is how quickly that opinion changes that really matters: "Everybody has to make some sort... of idea about somebody at first and it's how fast you change those that matters" (Audio, 4/30/08). Although Sophie did not voice agreement with the idea that race was a major factor in inequity, she reported being more attentive and deliberate in observing issues surrounding race in her classes. However, she seemed unsure of what to do, voicing frustration instead:

> I have started to observe race more in my advocacy class. I even had a very frustrated discussion with one AA [African American] male. I was frustrated with his behavior and the fact that he was focused more on his friends. I was frustrated because I know the AA males are scoring lower than they should be. (Written reflection, 4/20/08)

Persistent Conflict

By the end of the study, Sophie still appeared to be caught in conflict. She continued to employ strategies to discount race. She asserted positive values while still holding some deficit views. She was unable to accept race as the main component in inequity and instead continued to suggest alternate explanations. This limited understanding of the impact of race on equity kept Sophie from fully implementing practices that met the needs of her students of color. Sophie reported that the PD helped her in thinking about teaching rather than supporting her in making actual changes in practice, although she showed signs of movement in her reorganization of the socially just teaching of mathematics. Nonetheless, Sophie did not move out of the conflict she was experiencing between her teaching ideals and her deficit notions.

Sophie entered the PD with conflicts between a "socially just" philosophy of education and the dominant deficit scripts about race and learning mathematics. While she wanted to be a "socially just educator," her deficit notions continually created barriers that challenged her progression. One factor that limited her movement on the continuum was difficulty in recognizing race as a key component of inequity. Although throughout the PD, I noticed when teachers were not able to acknowledge a critical cause of the inequities in the schooling for students of color, trying to address the problem was difficult.

Discussion of Sophie's Journey

Sophie's journey throughout the PD characterizes that of many teachers who identify themselves as "socially just" educators yet struggle to effectively reject deficit notions. This conflict of narratives is highlighted by the work of Delgado and Stefancic (1992). They suggest that issues like racism can influence a personal "dominant narrative," an understanding and acceptance of the basic principles that form the foundation of how we reason. In Sophie's case, deficit notions of how students of color learn mathematics shaped her dominant narrative. She, however, also had a competing socially just narrative. This conflict was an instance of a new narrative being interpreted in light of old ones (Delgado & Stefancic, 1992). I argue that Sophie exhibited conflict in her ideas about students, and this conflict was a result of competing narratives, specifically the dominant narrative and the opposing newer narrative. The dominant or meta-narrative that framed her thinking was guided by racist and deficit notions and the new, less dominant narrative she was trying to adopt was informed by more progressive, "socially just" ideas. For Sophie, this conflict manifested itself in commentary on student learning as well as in repeated expressions of her deficit thinking. She would often offer alternative explanations for the existence of inequities and reject a focus on race.

Howard (2003) argues that a rejection of deficit-based thought about culturally diverse students is a central tenet of culturally relevant teaching. Sophie battled to dismiss these notions, and she continued to struggle to adopt the tenets of culturally relevant teaching. In Sophie's case, almost all of the artifacts, including readings, videos, and analysis of student work, failed to help her successfully move out of the conflict between her narratives. In fact, in some instances, the interaction with the artifacts seemed to reinforce her deficit notions.

Sophie's conflict was manifested in an avoidance and rationalization of racism. These rationalizations deflect the necessary focus from race in a way that is similar to how Ladson-Billings & Tate (1995) position multicultural education as raising issues of gender and sexuality to the same level as race. Sophie displayed what they describe as common "oppressor rationalization," "[where the] dominant group justifies its power with stories—stock explanations—that construct reality in ways to maintain their privilege. Thus, oppression is rationalized, causing little self-examination by the oppressor" (pp. 11–12). Sophie was unable to examine her perspectives and continued to believe the "stock stories" that framed her meta-narrative.

There was, however, at least one occasion in which Sophie rose to the challenge of reassessing her perspectives. An article by Birky and colleagues (2008) highlighted the successful journey of a teacher of African American students. The detailed examples of this teacher's instructional choices and narrative initiated a discussion within the group of teachers. This explicit example coupled with the ongoing assessment of her own practices through the analysis of student work created an opportunity for her to consider her practices. Sophie's assessment of practices focused on how students exhibited understanding and the methods she incorporated. Although these were not explicitly about issues of race, they segued into discussions of how racism informs teacher perceptions and the deficit notions that emerge in practice.

Implications for Future Research and Practice

The analysis of how artifacts facilitated and failed to promote movement helped me to recognize the complexity and challenges that arise when helping teachers discuss and reflect on race within teaching mathematics. This suggests that the needs of teachers can be extremely divergent.

Teachers such as Sophie who struggled to reject deficit notions presented the greatest challenges in supporting growth. Therefore, I am interested in investigating further how to better engage such teachers. Sophie's participation in and journey through the PD raise important issues around how to meet the needs of teachers who identify themselves as progressive and socially just in their practices and at the same time give evidence of deficit views. Given the lack of success the artifacts implemented in this PD had with teachers like Sophie who had a conflict between their meta-narrative and less dominant narrative, are there other artifacts that would be more effective in engaging them in unpacking issues of race and equity in mathematics education? Finally, to use artifacts as opportunities to engage teachers in discussions and reflections, I am interested in looking for ways to develop collaboration structures that can be sustained after the PD.

Note

1. Pseudonyms are used throughout the chapter, for both schools and individuals, to preserve anonymity.

Commentary: Part I

KYNDALL BROWN AND MEGAN FRANKE

The comments we offer here are our reflections on the salient issues that were raised in each of the chapters. We base these reflections upon our experiences as mathematics teachers, professional development providers, and researchers. It is clear that each of the researchers built on both theory and research to make sense of mathematics professional development focused on issues of equity. What is noteworthy is that the authors all make equity explicit, but they do not make it explicit in the same way. It is also evident that each of the chapters takes seriously what it would mean to engage teachers of mathematics in examining issues of equity in ways that challenge their existing practice. The authors draw on a range of theories and a broad research base; taken together, the resources themselves, provided across chapters, make a significant contribution to the field.

The chapters in the first section of the book highlight issues of teacher participation: who is participating and what learning trajectories are emerging as a result of the equity-based mathematics professional development. Each of the chapters provided details of the professional development practice and explicitly identified what the authors found to be the critical features of the professional development. In doing so, the authors closely attended to who participated within the professional development, how they participated, and how the participation shaped the professional development interactions. Here, we address three particular themes in relation to these ideas: noting teacher trajectory, negotiating and renegotiating relationships, and designing interactions.

Noting Teacher Trajectories

The learning and understanding that develops around issues of equity is complex and seemingly so different for individual teachers. The authors in these chapters describe what happens for teachers over time, and in doing so, provide a beginning outline of some initial teacher trajectories. These trajectories—of teacher knowledge, perspective, identity, and practice—are described in terms of shifts as well as sequences of shifts. Ho, particularly, begins to explicate a trajectory that details the shifts she saw in relation to teachers' accepting or challenging the dominant narratives of why students of color are not always successful in school. She examined the conflicts teachers experienced at different points in their learning. She then considered what type of support enabled teachers to work through the conflict they were experiencing.

In many ways this same approach is seen throughout the chapters in this section. While each chapter addresses the particulars of what artifacts, tools, and forms of engagement support teacher learning about mathematical equity, they each do this in relation to shifts in teacher learning specific to their group of teachers. As is the case with much professional development, the teachers described in these chapters arrived in different spaces with respect to their relationship to equity in mathematics. Wager reported on teachers who were at different starting points in their trajectories to develop a culturally relevant pedagogy. Celedón-Pattichis and her colleagues worked with teachers who came to the professional development with knowledge about the linguistic and cultural backgrounds of their students and concerns about how to meet their needs. In Foote's chapter, the teachers were willing to gather information from the parents of their students, but only one teacher was comfortable using something she learned to alter her curriculum. Rubel described teachers who were willing to try culturally relevant tasks in the classroom, but were reluctant to interact with members of the communities in which their students lived. We often recognize the complexity of supporting teachers to challenge their assumptions related to marginalized students and the reasons they may not succeed. What is noteworthy is that the professional development efforts reported on here were often successful in moving teachers beyond their initial ideas. The professional developers found ways to support teachers to consider what issues of equity in mathematics mean for their practice. Similarly, the authors engaged with teachers who came having already begun the process, encouraging them into the next phase of implementation.

In each of these chapters, the professional development providers recognized where teachers were in their own trajectories, and leveraged the strengths that teachers brought to move toward equity-based mathematics teaching. They created structures and drew on artifacts that allowed teachers to participate in ways that pushed their thinking forward with respect to equity pedagogy.

We need to understand that everyone is not going to move along that trajectory at the same pace. Different ideas are going to resonate at different points in people's lives, at different times, and in different contexts. The focus of our professional development work becomes about helping teachers to make progress. What is critical is the process of engaging, not simply a focus on the outcome of changing teacher practice. We want teachers to see themselves as always engaged in the process of learning about students and about their students' cultures, language, and race. We also want them to continually reflect upon how that helps them meet students' needs in the context of classrooms.

Negotiating and Renegotiating Relationships

The researchers in these chapters took what Edwards named as a non-deficit approach to the teachers engaged in their professional development. Taking a non-deficit approach requires understanding who teachers are and what they bring, and this requires getting to know teachers and continually building relationships with them. We believe that all of us who engage with teachers in professional development would agree that both getting to know the teachers and having them know us is important to the success of the professional development. The chapters in this volume, however, challenge us to think differently about what it means to get to know teachers.

The researchers recognize that each of us comes to the work of mathematics and equity with vastly different experiences, experiences that have informed our knowledge, our perspectives, and our views of ourselves as mathematics teachers. They have shaped our identities. These experiences are not ones to be explained away. They are to be understood, respected, discussed, countered with alternative evidence, and contributed to with new experiences. Addressing these experiences and identities requires getting to know each other in ways that are beyond what we know about mathematics, what we might typically do in our classrooms, and such. We need to get to know each other as people, see each other's histories as well as strengths and struggles. This takes ongoing attention, persistence, and time. We do not get to know each other after a single meeting or professional development session. We do not necessarily know each other in these ways because we teach in the school or have worked together before. We need opportunities to continually learn about each other and we need to make space to have this happen over time within the professional development setting.

The field has created a lot of structures around doing mathematics together in professional development that have allowed for a range of understandings to be shared. The authors in this volume extend traditional opportunities for teachers to learn about each other, opportunities often embedded in the ongoing work of the group. The professional developers allow space for the formal conversations that appear within these activities to continue. They also allow the informal opportunities to emerge. They extend informal conversations before and after the professional development, encourage the conversations that occur outside the professional development setting, and invest in bringing the conversations outside the professional development among the teachers into the professional development. The work of the professional developer around informal conversations is purposeful, aiming to unpack, uncover, come to understand teachers, and have them understand each other.

Who the professional developers were mattered. Their histories, experiences, knowledge, perspectives, and identities shaped how they designed, adapted, and engaged in their professional development. We realized as we reflected upon this that we know little about the *professional developers'* identities from the work presented here. We do think as we share about this work, those identities need also to be shared if we are to understand what shaped the professional development and it's outcomes. We wonder what professional development providers make of the knowledge, skills, and identities they bring to the professional development session. How do those elements shape the work they do? To what extent do race, ethnicity, and gender play in the minds of the teachers who attend the professional development? What difference does it make when the professional development provider has the same or a different background from the teachers? What difference does it make when the professional development provider has the same or a different background from the communities in which the teachers teach? We feel that the answers to these questions will help us to think more deeply about what is important when designing equity-based mathematics professional development.

COMMENTARY

Designing Interaction

One of the most significant contributions of the chapters is the attention to the particulars of the professional development. The authors have created or adapted tasks that help teachers grapple with issues of equity in mathematics professional development. The authors attend to what activities, tasks, and artifacts are used to structure opportunities, and what kind of learning takes place. Yet, it is not the activities, tasks, and artifacts in and of themselves that they shared. Rather, they shared *how* those activities, tasks, and artifacts operated in relation to the context, the teachers, and the content.

Wager's approach utilized dialogue structures and articles to engage teachers in discussions about equity. Koehn posed social justice math tasks to get teachers to struggle with what it means to teach mathematics for social justice. Edwards incorporated the use of video. Rubel used a mapping task as a way for teachers to learn more about the communities in which they were teaching. Foote drew on photographs. However, it is clear in all cases that it is not simply the particular artifact, it is the details of the artifact itself, and the way it emerged and was used in the context of the professional development, that shaped teacher participation and learning.

In Wager's study, teachers were resistant to implementing culturally based or social justice math lessons. The use of dialogue supported teachers to get issues on the table and interrogate their assumptions about equity and social justice. Foote chronicled the ways teachers' utilized information gathered from students to guide their instruction. Traditional ways of engaging with parents are usually confined to parent conference types of meetings. Using photography as a way to explore differently student understanding of mathematics redefined the relationship between teachers and parents because teachers had to rely on the knowledge that parents had about their own children, as opposed to the assumptions they made on the basis of classroom interactions. Koehn deliberately tried to create opportunities for struggle. The artifacts she chose, in one case the social justice mathematics tasks, created opportunity for struggle and for struggle that allowed for varied participation over time.

In each case the exercises, tasks, and artifacts allowed teachers to dive beyond the surface features and detail the particulars. These artifacts created focus consistent with a set of principled ideas at the basis of the professional development. In each case teachers participated differently and evolved differently in relation to the particular exercise, task, or artifact. What became salient and useful was related to who the teachers are, where they are in their trajectory of learning about mathematics and equity, and what the situation in which they find themselves is. How artifacts supported teachers depended on the teachers themselves, the context of the professional development as well as the context of the classrooms and schools in which the teachers participated.

PART II

*What Tools Have Proved Effective?:
Examining Effective Resources and Vehicles
for Addressing the Needs of Teachers
and Their Students*

CHAPTER 8

Building Community and Relationships that Support Critical Conversations on Race: The Case of Cognitively Guided Instruction

DAN BATTEY AND ANGELA CHAN

The focus of mathematics professional development for elementary teachers has shifted over time. In the case of Cognitively Guided Instruction (CGI), it started with building a cognitive framework of the development of children's thinking and carried this cognitive perspective into work with teachers (Carpenter, Fennema, & Franke, 1996). Over the years, some scholars have shifted toward a more situated perspective in working with teachers (Franke, & Kazemi, 2001). The focus of this work has moved from seeing individual teacher change to change in communities, relational identities, and generative growth (Franke, Carpenter, Levi, & Fennema, 2001; Franke, Kazemi, Shih, Biagetti, & Battey, 2005; Battey & Franke, 2008). More recently we have been searching for a way to build upon our work with school communities to foreground issues of race and equity. We continue our focus on children's mathematical thinking, but more explicitly address the deficit ideology that can remain despite the amount that has been achieved in this professional development work (Franke 2009; Battey & Franke, 2009). This chapter describes our progression from a focus on students' mathematical thinking to engaging teachers in grappling with the complexities of race in their classrooms. We leverage critical concepts in our earlier work such as community, relationships, and evidence to explore such conversations with teachers.

We begin by discussing our CGI work broadly, then focus specifically on how notions of community and relationships are critical to this work. From there we look at ways we have tried to challenge deficit ideas about African American students that have been raised in professional development. While this has produced individual teacher change, we think that we must go further. In the final section we discuss our attempts to bring issues of equity and race raised with individual teachers to the community level.

In writing this chapter, we are not arguing that previous work was limited or misguided. Instead, we have found that previous work has been critical in setting a foundation for pursuing difficult conversations around race. Understanding identity, relationships, and community are essential in dealing with the complex issues of equity that are raised within classrooms.

Cognitively Guided Instruction Overview

CGI is a professional development program, built around the principle that students bring intuitive knowledge and understanding to problem-solving situations (Carpenter, Fennema, Franke, Levi, & Empson, 1999). Within CGI professional development, facilitators share research-based knowledge of the kinds of strategies young children use to solve whole number operation problems. This professional development is structured around children's mathematical thinking, exploring the nuances of what makes problem types different from each other and why children may use particular strategies to approach the various problems. Student strategies are portrayed as following a developmental trajectory, beginning with concrete ways of modeling the action in problems and logically building toward more abstract strategies that are connected with larger mathematical ideas. CGI professional development engages teachers around the details of children's mathematical thinking, using artifacts of practice such as the student thinking, identifying what mathematical understanding students currently demonstrate, and building on and leveraging that understanding to move toward more sophisticated mathematical ideas.

Engagement around CGI in professional development settings has taken many forms over the years. Within professional development workgroup meetings, we engage teachers around mathematics tasks, artifacts of practice, and classroom experiences (Fennema et al., 1996; Kazemi & Franke, 2004; Franke et al., 2005). Teachers learn about children's mathematical thinking and begin to grapple with ways they can elicit their own students' thinking; within professional development we support them to consider elements of classroom practice that assist students in articulating their thinking and leverage that thinking to create robust learning opportunities. As teachers begin to try out different ways of engaging with students around mathematics, they pay close attention to the details of the knowledge their students have. Within professional development settings, we help teachers to build on students' mathematical learning by capitalizing on what students *can* do rather than focusing on what they *cannot* do, and support further understanding in ways that are connected to what students know (Jacobs, Franke, Carpenter, Levi, & Battey, 2007).

As teachers begin to use CGI ideas to guide their teaching, a noticeable shift in their practice is the level to which they pay attention to *each* student in their classroom. Teachers become skilled at knowing their own students' mathematical thinking and what that indicates about their mathematical understanding. The work on children's mathematical thinking in professional development pushes teachers to move past sweeping generalizations of who is and who is not good at mathematics, focusing purposefully on the evidence of mathematical understanding that students demonstrate. The artifacts of practice that teachers bring to professional development meetings allow for a grounded way to discuss students, connecting teachers' claims about what students know to specific details. This extends to teachers' stories about what students can and cannot do, providing productive spaces for professional development facilitators to challenge deficit notions and address critical issues of race and equity. Before this can happen, however, a strong community foundation must first be established.

Developing Community and Relationships around Student Thinking

One of the guiding principles driving our work with schools around CGI is the development of community and teacher identity, and the fostering of relationships. Theoretically, we draw from a broad range of work that informs our notions of supporting a school community to pay close attention to student thinking. We draw on the works of Rogoff (1997) and Wenger (1998) in seeing teaching as more than the sum of knowledge, beliefs, and skill, instead viewing it as occurring through participation in a community of practice (Battey & Franke,

2008). Therefore, teacher learning is contingent on the conception of community. Community is contingent on dealing with the fault lines of practice (Grossman, Wineburg, & Woolworth, 2001), the "faces of practice" that are made visible or not (Little, 2002), and the cultural tools that mediate learning (Boreham & Morgan, 2004). Community and teacher learning are relational concepts based on

> the histories, structures, as well as colleagues, providing contexts for teachers to construct possible identities, and allow new teachers the opportunity to "practice their identity" (see Holland, Lachicotte, Skinner, & Cain, 2001), for full membership in the community of practice. In this sense, the contexts in which we participate guide us in developing who we are. Contexts can constrain or open up new possibilities as other teachers practice mathematics instruction in particular ways, the curriculum embodies a particular take on what mathematics is, and students bring their own notions of what it means to do mathematics. (Battey & Franke, 2008, p. 129)

These relations can be viewed through classrooms as teachers practice their identities, engaging with students and the content. Teachers enact identities as they engage in practice, broadly conceptualized here as including professional development and the classroom.

In addition to the more "formal" professional development workgroup meetings, a critical element of our model of professional development is the focus on the more "informal" interactions that occurred during onsite support (Kazemi, 2008; Franke & Chan, 2009). Spending one half-day per week at the school, we provide onsite support as teachers figure out how to connect CGI ideas from professional development settings into their daily teaching repertoire. We spend time in classrooms, talking to students while they work on mathematics problems; jump in while a teacher is asking questions of students; and converse with teachers in the hallways and lounge areas. Being present at the school allows us to provide "real-time" support to teachers, fielding questions as teachers are just about to try something new and debriefing with them about their experiences and possible next steps.

Although onsite support has been an important piece of our professional development work for years, we are highlighting it here and identifying it as the core of our engagement with teachers (Fennema et al., 1996; Franke et al., 2005; Jacobs et al., 2007; Kazemi, 2008). Our interactions have become much more purposeful; teachers know exactly which day of the week to expect us, and we make sure to check in with teachers, visit classrooms upon request, and "hang out" in the hallways and staff lounge. We want to have our presence known on our days at the school so that we become an expected part of the weekly school routine. We emphasize the "informal" nature of the onsite support to highlight how it allows us to get to know teachers and develop relationships with them. Our role at the school is not to, say, model demonstration lessons, but to become an ongoing part of a school community that is committed to increasing student learning, paying attention to student thinking, and acquiring some knowledge and skill.

As we become a part of the school community, we develop connections with individual teachers and simply get to know them: who they are, what their mathematics teaching is about, how they see themselves in relation to their teaching. In this sense, teachers' stories are critical—how a teacher thinks of herself in relation to "a particular context, with a particular history, with others who have ideas about themselves. These histories (and the structures in which they are embedded) contribute to how a teacher comes to make sense of what it means for her or him to be a teacher" (Battey & Franke, 2008, p. 128). We tune into areas in which teachers see themselves as particularly skilled and areas in which they exhibit more hesitancy. Similar to how CGI teachers would engage with students, we capitalize on teachers' already existing knowledge and strengths and leverage them to push their learning. We build relationships with teachers in a way that connects to who they are as teachers and their classroom practice, establishing a foundation around which to further their teaching.

We also use what we know to be individual teachers' strengths to foster collaboration among teachers. Although some schools we work with have already developed a strong sense of community, we push further to support relationship building specifically connected to classroom practice. As we engage with teachers onsite, we informally make explicit particular ways they could collaborate around CGI ideas. We highlight which teachers have developed interesting problems that could be shared with fellow grade-level colleagues and which teachers have figured out how to connect warm-up activities to problems. We support conversations among teachers who have already developed a routine of trying things out together. During onsite support and within formal professional development settings, we position teachers as skilled and encourage them to share their strengths with each other.

Washington Learning Community: A School "Doing CGI"[1]

Throughout the first year of our work at Washington Learning Community, teachers grappled in different ways with the idea of "doing CGI." As professional development facilitators, we never explicitly used this phrase because CGI is not a make-it-and-take-it activity and there are a multitude of different ways to incorporate CGI principles into one's teaching. However, the language of "doing CGI" emerged among the teaching community and indicated that teachers had particular notions of mathematics teaching in relation to eliciting and building upon students' thinking. For teachers at Washington, "doing CGI" meant posing word problems, encouraging students to use different strategies and different mathematical tools, and supporting students to share their strategies in detail. Teachers saw themselves as either doing CGI or not doing it, and this varied among teachers and as the year progressed. Some had already seen CGI as an integral part of their mathematics teaching, while others participated more peripherally with CGI. Some teachers tried out new ideas from the beginning, while others took more time to figure out how to bring the professional development ideas into their classroom.

By the end of the first year of our work, there was a clear shift in the teaching community in relation to "doing CGI." Though participation with the professional development varied across teachers and in classroom implementation, it was apparent that Washington had become a school community that was using children's mathematical thinking to guide instruction. In classrooms, teachers were skillfully engaging students around math problems, supporting them to articulate their own strategies, asking questions that highlighted mathematical ideas, and connecting new mathematics content to what students already knew. The school administration recognized their mathematics program as guided by CGI; outsiders looked to Washington for examples of classrooms where instruction was guided by CGI principles. The teaching and learning of mathematics at Washington had clearly shifted as teachers paid attention to individual students, created environments that met individual students' specific needs, and supported different kinds of participation in relation to mathematics.

As our work in schools continued, we developed stronger relationships with more and more teachers; we engaged around stories of their students' mathematical thinking and we attempted to figure out what they wanted to try next to push mathematical understanding. These stories continued as teachers became more skilled at paying close attention to individual students and shared evidence of their mathematical thinking. Over time, our relationships with teachers, grounded in their shifting classroom practice, created a foundation on which we could address more critical issues. Building upon this growth, we began to more explicitly deal with issues of equity that play out in classrooms, specifically countering deficit notions of students of color.

Challenging teachers' deficit notions at the individual level

In our work with teachers, we often hear stories about students of color. The stories told about African Americans and Latinos often relate to *metanarratives,* or broad stories, passed on through the media about lack of parental involvement, not valuing educational achievement, and an innate lack of mathematical intelligence (Giroux, Lanskshear, McLaren, & Peters, 1996). Within mathematics these metanarratives present African Americans as inferior to Whites and Asian Americans; failure is considered normative, and the struggle to obtain mathematics literacy despite barriers and obstacles, receives little or no attention (Martin, 2007). Despite evidence and statistics to the contrary, the everyday stories told on television, in the movies, in daily conversations, and in newspapers perpetuate powerful ways of framing students of color within society. These broad stories contain ideologies within them, ideologies that recreate stereotypes of the "other." They structure the ways of talking about children of color, communities of color, and frame our individual actions (DiME, 2007).

In our professional development work we have begun to counter the metanarratives about the mathematics that students of color *can't* do. Shifting our lens to a perspective that sees teaching as participating in a certain kind of discursive practice (beyond the local community) allows an understanding of knowledge, beliefs, and skills as embedded within the broader narratives that teachers engage in during everyday practice. It also helps us understand why many teachers have deficit views of students and how these ideologies are passed across teachers and schools. A metanarrative can construct how to organize students' abilities in the classroom, for instance, as high, medium, and low.

As teachers participate in professional development they gain new knowledge, skills, ways of talking, and teaching practices—new ideologies—and these new tools facilitate a shift in identity—a shift in what it means to be a mathematics teacher. Therefore, as teachers participate in narratives related to mathematics, mathematical discourses, they are not only sharing their knowledge, skills, and stance on what it means to teach mathematics, they are also reproducing narratives about *who can do mathematics*. The question then for professional developers becomes: Can we construct ways of challenging these discourses within local school contexts as a way to open access for students of color to engage in substantive mathematics? We are arguing here that a foundation of focusing on evidence of student thinking provides a space for professional developers to do so.

As teachers attended to students' mathematical thinking, their teaching shifted to be more tuned in to the needs of individual students. Paying close attention to the details of students' mathematical thinking allowed teachers to notice different patterns of participation in their classrooms, connected to specific evidence rather than broader narratives of who can and cannot do mathematics. Working with teachers who share evidence about their students' thinking in a detailed manner allowed for us as professional developers to engage with them in particular ways.

Supporting struggling students at Washington Learning Community. Our time in teachers' classrooms allowed us to gain insight into student participation, knowledge about the teachers' practices, and space to have conversations with teachers about vulnerable issues. A few teachers began to come to us with concerns about some of their students' participation, specifically their African American students.

One second grade teacher at the school approached us in the hallway during the school day and said that she was worried because her three least successful students were all African Americans. "They just have so few skills and they're so far behind.... I'm worried that they won't catch up and will always be behind." This teacher noticed this issue in her classroom because she was paying attention to individual students and the details of their thinking. Though we had been discussing student thinking, focusing on what students *can* do, and finding ways to help students succeed, this was an instance of deficit ideas being placed on African American children with little evidence of their mathematical understanding being shared. It was also an opportunity to have a meaningful discussion with a teacher in a very vulnerable moment.

In response, we asked the teacher questions about what her African American students specifically *could* do: Can your kids count by ones? Tens? Can they solve single-digit number facts? Can they solve Join Result Unknown story problems? As the teacher answered these questions specific to our work on children's mathematical thinking, she acknowledged what her students *could* do and realized her students did

Building Community and Relationships

have counting skills in particular that she could build upon. Within our conversation, she planned what she could do in class to support these students.

Rather than latching onto a negative broad narrative of student failure, we asked questions that redirected the teacher to the details about what the African American students did know. As the year went on, the teacher created more opportunities for the three African American students to share, showing her and the other students that these children were mathematically capable. As the year ended, the three students were in the middle of the class in terms of their mathematics knowledge. By shifting the focus from a negative metanarrative citing African American failure in mathematics, and relying instead on the knowledge and skills that the children already had, the teacher was able to develop problems and create opportunities to allow the students to interact and display the ability they did have. Through an informal interaction in the hallway during the school day, the teacher's focus was shifted, and with continued support in facilitating participation, opportunities were opened for the African American students to build mathematical identities and the teacher to see them as mathematically able.

This example highlights our work with one teacher, reframing her notions of how to support her African American students. Our efforts with other teachers were individualized as well, not in the sense that we treated each teacher as an isolated entity, but that we thoughtfully considered each teacher's individual history in our interactions with the teachers. How we engaged with each teacher differed slightly—in the issues we discussed, the kinds of support we provided for their teaching, and the ways we participated in classrooms with their students—on the basis of who each teacher was and what each brought to our joint work. What was consistent for us across the school, however, was our commitment to challenging deficit notions of students of color, particularly the African American students that the school was having difficulty serving, and drawing upon classroom experiences and providing evidence to highlight what students *can* do.

As our work with individual teachers progressed, we pushed ourselves to foreground critical issues at the community level. On these sensitive topics, teachers approached us individually while we were in their classrooms or informally in the hallway. We argue that the relationships we had established with teachers around their students and classroom practice allowed for these conversations to happen. Having these more nuanced conversations informally and within the walls of the classrooms, however, did not mean that they automatically surfaced across the larger school community, particularly within more formal professional development settings. We characterize this somewhat as isolated teacher change around issues of equity and race within broader community change around teaching mathematics for understanding. However, we considered it critical that conversations like this one reach the entire community; so we began to explore more overt discussions of race at the community level.

Addressing Deficit Notions at the Community Level

In shifting the conversation to the community level, three issues were critical. First, we wanted to maintain the centrality of the mathematics in addition to attending to issues of equity. Second, the issues were raised *by teachers*. Rather than us coming in with our own ideas about how these conversations would progress, teachers raised issues of race from their own classroom practice. In this sense, discussions were grounded in teachers' practice and evidence of student thinking, practical and meaningful. The third issue was the amount of time we spent in teachers' classrooms. Though outsiders as university faculty and graduate students, the time we spent in classrooms gave us detailed understandings of the practices teachers were engaging in as well as the students in their classrooms.

Explicit conversations at Washington Learning Community. We acknowledged the community for the gains they had made in their teaching practice, in their collaborative efforts, in how their students could articulate their thinking, in their demonstration of conceptual understanding, and in their test score gains. This conversation became an entry point to discuss, amidst all of these gains, where teachers still felt they needed to progress as a school community: in supporting struggling students of color. We asked teachers to think about who their struggling students were and what kinds of labels were more globally assigned to them. Teachers identified labels such as "minority students," "English Language Learners," "not having family support," "children of color," "lacking number sense," "have parents without the tools to help them," "not willing to try," "domestic problems," and "those who don't verbalize anything." One of the issues that came out of this conversation was that the students this school was not serving well were African American students. While a painful conversation, it exposed the ways in which African American students are often framed in schooling; it made the deficit metanarratives accessed by teachers explicit. This was a beginning toward our first goal of having teachers question their own assumptions of who their students were: their culture, lives, and abilities.

We then made a shift with the community of teachers that we had been making with teachers on a more individual basis: We asked questions that shifted the framing of their notions about African American students from a deficit perspective of what they *couldn't* do to a more positive perspective focused on what students *were* doing, and more critically, *why*. From our work with teachers in their classrooms, we knew the specifics of what teachers were most worried about, such as the kinds of strategies struggling students used. We voiced these specifics in the whole group setting (i.e., second grade students using "direct modeling by ones," strategies that teachers consider too unsophisticated or concrete for their level) and asked teachers to inquire into *why* students used these particular strategies. Does the student know this is a strategy that will always work for them, even if the problem type is unfamiliar? Did the student hear other students receiving praise for using the same strategy? Is the student most comfortable explaining this strategy over others in case she is asked to share in front of the entire class? In asking these questions, we wanted to move teachers away from the common assumption that students of color use less sophisticated or less abstract strategies because they are unable to comprehend the mathematics within more sophisticated ones. Rather, we wanted to direct teachers toward the agency their African American students have and the purposeful decision making that underlies their actions: Students participate in mathematics in particular ways for a reason.

In addition, we wanted to root teachers' claims about students in evidence rather than assumptions. Instead of relying on deficit ideas about why students of color do what they do, we wanted them to get to know their students better and rely on talking to students. We asked teachers to grapple with this question: What experiences have African American students had that would lead them to choose particular ways to participate in mathematics? We wanted teachers to move away from treating students along the lines of the deficit narratives and respond to the nuances and complexities that do not fit with stock stories provided by society. It was critical for us that teachers get to know their African American students better and understand the choices students were making as contextual, responding to the immediate environment of the classroom considering prior experiences and histories.

In subsequent conversations we put another question on the table: Knowing that there are larger structural issues at play here, how do we, here in school and in our classrooms, support African American students in their mathematics learning? The teachers with whom we had more explicit conversations, such as the one we described in an earlier section, could take the risk to be vulnerable and raise further issues from their own practice. Our work with teachers in their classrooms positioned them to share stories about the ways they struggled to support African American students, the kinds of changes they made in their teaching to support them, and the continued progress they were making in their classrooms. This conversation created space for success stories to emerge, stories that stood in stark contrast to prevailing deficit narratives. It also created space for teachers to continue to grapple with how they could better support marginalized students.

For us, this is a way of addressing Gay's (2002) assertion that "because culture strongly influences the attitudes, values, and behaviors that students and teachers bring to the instructional process, it has to likewise be a major determinant of how the problems of underachievement are solved" (p. 114). We leveraged the work that we had done at the school over the course of three years to address determining factors that influenced the continuing struggle of students of color, building on the relationships we had developed, what we had accomplished with individual teachers, and staying closely connected to the local school community.

Conclusions: Critical Themes Highlighted in This Work

While we see ourselves as continuing to learn to do this work and searching for new avenues to have these conversations with teachers, we argue that the central constructs in prior work bring us closer to addressing deficit notions of race in an explicit, authentic, and practice-oriented way. From the original CGI professional development, to gathering evidence of student thinking, to issues of community and identity, to countering deficit ideas, each phase of prior work is a building block for having conversations about race, allowing each conversation to emerge more explicit than the next. We think that building a stronger community in which teachers are not isolated, centered on a norm of making practice public, is crucial.

Our relationships with teachers as well as the time we spend in classrooms allow for more vulnerable conversations to be had. In addition, we are building knowledge of classroom practices, the local culture, and students, as we become more central participants in the school community. Our shifting of attention to what students *can* do allows for some of our relationships to begin to address issues of student participation, success, and classroom practices that support the achievement of students of color.

Finally, recognizing the successes and struggles of the teachers and the school allows for continued growth. Early conversations with individual teachers about dealing with participation in their classrooms allow for issues of equity to be raised by teachers. In highlighting the critical nature of this piece, we are challenging some of the research community to value the practical knowledge and experience of teachers in this work rather than coming in and telling teachers to grapple with issues of race. If teachers are not respecting students' knowledge and experiences, the solution is not to do the same to teachers. Instead, we value teachers' knowledge and experiences and use this as the impetus to push further on issues of race. In this way, teachers continue to learn, question, and build a stronger school community that can begin to deal with complex issues related to race, even after we leave the school.

As we continue this work, we realize we are just beginning to figure out how to integrate issues of race in professional development. Raising these issues at the school community level is not easy, particularly when schools have a history of not serving a particular student group. For us, conversation with teachers about race must be done in an authentic, practice-oriented, public way that respects teachers, just as we want teachers to respect students. Making deficit narratives explicit is not an easy task while building practical knowledge of teaching mathematics for understanding. But relationships, community, artifacts, and struggling with difficult issues of practice are critical for taking on risky, vulnerable, and contentious discussions of race.

Notes

The authors share equal authorship on this chapter.

1. This chapter draws on the work of Franke & Chan (2009) as they engaged an elementary school community in CGI over the course of three years.

CHAPTER 9

Status and Competence as Entry Points into Discussions of Equity in Mathematics Classrooms

VICTORIA M. HAND, JESSICA QUINDEL, AND INDIGO ESMONDE

Equity and Group Work in Mathematics Education

Students do not all benefit equally from participation in collaborative discussions. Issues of equity can be seen in mathematical group work in a variety of ways if equity is defined as a *fair distribution of opportunities to learn* (Esmonde, 2009). Studies investigating how equity issues arise have typically focused on whether students have access to correct mathematical explanations.

If equity is defined in this way, then one might argue for the benefits of heterogeneous grouping in which high-achieving students and low-achieving students are placed together. In such groups, the high-achieving students would be positioned by the teacher (and possibly, peers) as the experts in charge of helping the other students figure out what to do. Or, one might argue for homogeneous "ability groups" in hopes of tailoring instruction to students of differing achievement levels.

However, access to correct explanations does not guarantee that students will learn mathematics. Theory and research in mathematics education suggest that joint interaction with peers actually contributes more to learning than mere exposure to correct answers (Barron, 2003). After all, group work often takes place in the context of a number of other kinds of activities in the classroom, in which students will have the opportunity to see and hear correct explanations.

Therefore, in contrast to an approach to equity that focuses on access to explanations, we focus on whether students are able to interact meaningfully in cooperative groups. If the benefit of group work lies primarily in the interaction, then it is inequitable to have some students act as "experts" who do most of the talking, while other students act as "novice" learners who mainly listen, and whose ideas are not taken into account. Complex Instruction (CI), a pedagogical framework created by Cohen, Lotan, and their colleagues (Cohen, 1994; Cohen & Goodlad, 1994; Cohen & Lotan, 1997b), provides guidance for both understanding this problem and intervening to rectify it. They argue that in classroom interaction, students have differential status, and that these status differences are primarily associated with who is seen as academically "smart," and who is popular. High-status students dominate group discussions and also benefit the most from them. The authors suggest several interventions to reduce status differentials in the classroom.

They first recommend the use of "group-worthy problems" that require group members to coordinate their efforts around powerful mathematical ideas (Boaler, 2006a; Lotan, 2003). The teacher then implements the "multiple abilities treatment," explicitly telling students the multiple kinds of competence that are required to solve the problem, and reminding students that no one student has all these forms of competence, but all of them have some. Then, when students are working together on these tasks, the teacher observes their interactions and intervenes to "assign competence" to low-status individuals whose contributions are going unnoticed in their groups. When students have participated in classrooms using the CI approach, and in which teachers have made active efforts to equalize status, they develop what Boaler (2006b) calls "relational equity": "positive and respectful intellectual relations" (p. 78).

Defining equity in this way emphasizes the importance of students coming to respect one another, and value different forms of competence. However, we would caution that a colorblind definition such as this one might actually mask the ways in which classroom patterns of interaction reinforce racialized, gendered, and socioeconomic hierarchies. Taking steps to minimize status differentials in the classroom is even more critical because of the ways in which classroom status aligns with power, and who has power in a classroom often reflects the power structures in society (as we will discuss later in relation to the studies reported here). Creating classroom activities in which many students can be successful, and highlighting their success by assigning competence to students whose contributions are being ignored, has the potential to alter not only the way students see one another, but the way teachers see students. As students from nondominant groups come to be viewed as academically engaged and successful, negative stereotypes about these groups of students are contested.

Investigating aspects of CI in professional development settings, then, holds promise as one approach to helping teachers develop more equitable classroom practices. Change in teacher practice is not easy to come by, particularly when the changes advocated are ambiguous and complex (Ball & Cohen, 1999; Eisenhart et al., 1993). Currently, teachers are being asked to teach equitably and toward broader social justice goals (Gutstein & Peterson, 2005, Moje, 2007; North, 2006). However, myriad versions of what it means to teach for equity (Gutiérrez, 2002) have resulted in a menu of orientations and practices among which teachers must choose.

The complexity behind these choices is not always apparent. First, theories of equity within the field of education are largely underdeveloped, complex, and sometimes clash with one another (DiME, 2007). This ambiguity can be problematic particularly when discussions focus squarely on difficult topics such as race and privilege. Teachers can feel attacked, unfairly criticized, and thus vulnerable (Wilson & Berne, 1999), and even more so when asked to reflect on discrimination and social injustice (Rodriguez & Kitchen, 2005). Second, recent studies on professional development focused on equity in mathematics have found that it is difficult for teachers to reject a deficit perspective of lower-achieving students (Foote, 2006), to view social justice as more than simply an add-on (Gau, 2005), and to get past the cultural relevance of a task to the cultural relevance of a classroom (Hand, 2003). What we mean by the latter is that tasks embedded in contexts that are culturally relevant are often assumed to reach a wider range of learners, whether or not the classroom norms and practices are based on principles of cultural relevance as well. Finally, teachers are often told how to conceptualize issues in teaching and learning, instead of giving voice to the issues that arise in their own classrooms.

In this chapter, we report on two classroom-based research studies that investigated status-management techniques based on the CI approach to address issues of equity among students. Both studies were driven by teachers' desire to effect change in their classrooms that would begin to redress broader issues of social injustice.

Study 1: Managing Status as an Entrée to Teaching for Equity

In this first study, I (Hand) report on a *professional development working group* designed to provide teachers and researchers with opportunities to explore a range of approaches to equity in the context of teachers' classrooms. This working group found that conceiving of equity in terms of *status* became a productive means by which teachers could imagine and begin to generate more equitable participation among their students.

The working group model of professional development has been shown to be effective at disrupting predefined status hierarchies between the "expert" researcher and "novice" teacher (Fennema et al., 1996). Within working groups, teachers and researchers identify a problem to be worked on, and consider this issue in relation to the multiple contexts of teachers' work.

The professional development working group in this study focused on engaging in critical yet constructive conversations about equity in mathematics, as both a goal for their individual classroom instruction and a mechanism for repairing broader social injustice. These conversations were supported by records of practice that resulted from investigations carried out in the teachers' classrooms during the professional development (Franke, Kazemi, & Battey, 2007; Little, 2002; Wilson & Berne, 1999). The *status* of the students in the teachers' mathematics classrooms became a major focus of this group and a productive means for envisioning and creating broad-based participation among students.

Method

The working group consisted of six women, including two middle school mathematics teachers, a mathematics education professor, and mathematics education graduate students. One of the participants was a Latina; the rest were White. The approaches examined over the course of a semester included culturally relevant pedagogy (Ladson-Billings, 1995), CI (Cohen & Lotan, 1997a), Funds-of-Knowledge (Moll, Amanti, Neff, & Gonzalez, 1992), and teaching for social justice (Gutstein, 2006). Emphasis was placed on working with teachers to frame issues of equity in their mathematics classrooms and to agree on possible approaches to equitable instruction to investigate within their classrooms. The working group read and discussed articles on these approaches at monthly meetings, and experimented most extensively with CI in their classrooms. Teaching experiments around CI were videotaped, and the videotapes were analyzed during the monthly meetings.

Data collected included audiotapes of working group meetings and teacher interviews, videotapes of classroom experiments (whole class and student groups), and written reflections by teachers and researchers on the working group meetings. Analysis focused on the following research question: How does attention to issues of classroom status help teachers reframe the lack of participation among their underrepresented students? The audiotapes of the meetings, interviews, and participants' reflections were transcribed and coded. The videotapes were then examined in conjunction with key events in the primary transcripts.

Results

Due to the small number of participants in the working group, we consider the results of this study to be preliminary. As such, an important theme that emerged from this analysis is that the working group found the most purchase for equitable instruction came with the CI approach. To experiment with the CI approach, the group used a group-worthy mathematical task appropriate for sixth grade students, the Locker Problem, in combination with structured group work and the assignment of competence in two classrooms.

Managing status among students. The CI approach proved to be appealing to the teachers for several reasons. First, assigning competence to non-dominant students around significant mathematical ideas afforded immediate and visible changes in students' participation patterns. Teachers initially wondered how the issues being discussed in the working group (e.g., assigning competence, changes in status) would play out in their classrooms among all of their students, particularly students that teachers identified as having learning disabilities or behavior issues. However, in their classrooms, following the CI session, teachers noticed these students engaged in mathematical discourse within small and whole class discussions, which was further validated upon reviewing classroom videos. An example of this kind of insight is illustrated in the following episode, in which one of the teachers (T) is reflecting on her experience with Darrius,[1] after watching a clip of him during the working group meeting.

★★★

T: Darrius, the African American boy, he's special ed. He really can't read. That's his major problem. Peng does very well in math, so does Juan. Juan is bilingual, Spanish, and Peng also speaks Hmong. But they're both high level, you know. They assign...ESL assigns levels to them and they are like high levels, like four or five. They don't need ESL support.
Hand (H): How does Darrius generally participate in class?
T: He does not.
H: Okay.
T: Well he does if you're in a small group and you sit right next to him. This [CI session] really pulled Darrius out. I mean Darrius was like, I was like, "Wow, Darrius!" And at first I didn't know if it was the videotape. But even today without the videotape he was all, it was, it was really interesting to see who piped up and offered information in small group and then with the whole class...I'm not sure what it was, but status changed for a number of children doing it this way. He was one of them.

This brief excerpt points to the teacher's perspective of the significant shift in the patterns of participation for some of the students in her classroom, and in the role of status in prompting these shifts. In her initial utterance, the teacher singles out three students and positions them around particular institutional labels. These labels appeared to shape her expectations for the nature of the students' ability to contribute to and achieve in her classroom. For example, she links Darrius's inability to read to his status as a special education student and ultimately to his tendency not to participate relative to the other students. Two other students mentioned, both labeled ESL, exceed her expectations by achieving at a high level. It is interesting to note here that while the students are characterized as members of broader groups, these groups are defined in terms of educational anomalies (e.g., learning disability, limited English skills). In other words, students are perceived in terms of individual deficiencies.

The CI experiment, although limited in scope, had profound implications for how students like Darrius came to be viewed by the teachers—both in terms of individual ability and the relation of student participation to personal characteristics and classroom structures. Instead of seeing Darrius's label as learning disabled as dictating his performance, the teacher acknowledged that his classroom status was also playing a role. While the teacher did not go as far as to begin to notice patterns between racial and ethnic groups of students and status hierarchies and classroom contributions, her ability to see past Darrius's deficit label is important. The video clips provided a means of identifying the effects of status in the teachers' classrooms, and in coming to agreement on how changes in status changed student participation.

Implications

What are the implications of this study for professional development efforts geared toward building equitable teaching practices? One implication is if teachers have an opportunity to overcome stereotypes about individual students based on how they are labeled, they may be inspired to engage in deep restructuring of their practice. Seeing students' participation change before their very eyes had a profound effect on how these teachers came to see what was possible in their classrooms and among their students. The teacher in the transcript decided to experiment more fully with CI over the following school year. A second implication is that for teachers to even begin to consider approaches that require changes to their curriculum, teacher educators may need to first address how these approaches can exist alongside of or within content and standards requirements. It may be unwise to ask teachers to make changes without acknowledging the realities they face.

Teachers' understandings of issues of equity can take many forms, which necessarily shape the goals they have for equitable classroom instruction. The section that follows delves into the experiences of one of the authors of this chapter (a high school classroom mathematics teacher) as she drew on strategies from CI to disrupt what she perceived as traditional lines of power and status in her ethnically and racially diverse classroom. In it, she explains how she grappled with creating a socially just mathematics classroom.

Study 2: Assigning Competence to Disrupt Culture of Power in Mathematics Classroom

I (Quindel) am a high school mathematics teacher at Berkeley High. For the past five years, I have been exploring pedagogical practices and mathematical content that support greater social justice in my mathematics classroom. I have come to look at teaching mathematics for social justice from two angles:

1. *How* we teach mathematics, aiming to create an equitable classroom in which all students have access to reaching their fullest learning potential
2. In *what* problem contexts we teach mathematics (e.g., doing fewer problems related to food and sports and more problems related to access to college, prisons, war, the environment, racism, etc.)

My bigger goal as a mathematics educator has been to teach mathematics for social justice by using problems in the context of social justice issues as well as using social justice pedagogy. However, I found that writing and rewriting curriculum was extremely challenging while being a full-time teacher and that focusing on *how* I was teaching was a good start toward becoming a social justice mathematics educator.

During 2005–2007, I participated in a program called Inquiry Making Progress Across Communities of Teachers (IMPACT). Project IMPACT was launched in 2003 by faculty of the University of California, Berkeley Graduate School of Education, to support K-12 classroom teachers —especially novices—who chose to serve in low-performing public schools in the San Francisco Bay Area. Project participants gathered biweekly to explore their own questions about teaching, learning, and schooling. An external facilitator (usually a doctoral student) assisted the group, and participants received a stipend to recognize their investment of time and energy. Throughout the school year, participants convened quarterly as a regional network for intensive professional development and to share what they were learning.

I conducted two action research projects in my class over the course of two years. For this chapter, I report on the first, in which I employed the CI strategy *assigning competence* to rearrange status in my classroom.

Assigning Competence

Throughout the first year of Project IMPACT, our study group focused on the idea of being explicit about what we expected from our students. We recognized that in the "real world," people in power (White people, men, heterosexuals, people who speak English as a first language, etc.) are trained within the culture of power and therefore maintain their power/privilege by following the unspoken rules that enable them to succeed. Though they maintain this power, the fact that they benefited from a system in which their actions, culture, and life experiences led them to their success is often invisible to them. The way privilege operates is also invisible to most oppressed people, who are often blamed for their own lack of opportunity, which the culture of power created (Eubanks, Parish, & Smith, 1997). Thus, my goal with my inquiry project was to be more explicit about what it means to be good at mathematics and how excellence can be achieved in a mathematics classroom. My aim was to undermine the invisible culture of power that existed in my classroom, a culture that was created because of the society in which we live and that needed to be challenged directly if it were to change.

All of my classes are taught using the Interactive Mathematics Program (IMP) curriculum (Fendel, Resek, Alper, & Fraser, 1998), a reform-based integrated mathematics curriculum that emphasizes problem solving, problems based on real-world situations, group collaboration, and communicating about mathematics through written and oral presentations. I chose to conduct the intervention in my IMP 2 class because it was the most diverse in terms of race, socioeconomic status, grade level, and mathematical skills. The demographic makeup of the class was as follows: 13 African American students, eight White students, two Mexican American students, and four Asian American students. Girls made up two-thirds of the class, and there were three English Language Learners, four students with disabilities, and seven students identified as "gifted and talented." While I loved the diversity of the class, I found that inequitable achievement patterns appeared early on in the year. The particular pattern that disturbed me most was the underachievement of most of the African American students while most of the White and Asian American students achieved at a much higher level. The study explored the following research question: Will assigning competence and emphasizing multiple ways of being good at mathematics in my IMP 2 classroom help all students learn more mathematics and decrease inequitable status issues in the classroom?

Method

This project particularly focused on what it means to be a "good" mathematics student. I began the study by distributing a questionnaire in which I asked the students who they thought was good at mathematics within our class. As I had expected, White and Asian American students were disproportionately perceived as being good at mathematics.

To begin to challenge these conventional stereotypes, at the beginning of the second semester, I began a new routine in my classroom. Every day, I would post a "competency" along with the agenda on the board. Competencies ranged from "justifying" and "generalizing" to "questioning" and "explaining," and I chose them on the basis of my goals for the day's lesson. Sometimes the competencies were specific mathematical ideas, but often they were ways of learning mathematics. At the end of the lesson, I would announce which two to three students or which group "won" the competency that day, and award those students with extra credit worth about the same as a homework assignment. When I chose which students were awarded the competencies, I kept track of who had won how many times and I would try to choose different people or competencies that leaned toward students' strengths, especially those who had not yet been chosen. Throughout the unit, each student was awarded the competency at least one time. At the end of the eight-week unit, I distributed a second questionnaire, which again asked, "Who is good at math in this class?" These questionnaires as well as the students' grades were the main sources of data I used in this project.

Results

When I analyzed the questionnaires, I was disappointed to find that the status of the students in the class had not changed much. White and Asian American students still disproportionately had the highest status. However, I had personally felt a significant shift in the dynamics of the classroom, with more and more students engaged and participating on a regular basis, and often trying to win the competency. This led me to look deeper into my data. Although there were other factors that may have influenced the data—such as four struggling students leaving the school, building more trusting relationships with students later in the year, and individual interventions with different students—my results indicated that assigning competency had positively impacted the learning environment for many students.

Of the 24 students surveyed, 19 wanted competency rewarding to be continued and only three did not (two were indifferent). Two of these three students were not against the practice, but against *how* I assigned the competencies. I decided to act on the suggestion of the IMPACT group that I directly ask the students what they thought of the new routine. I found that this qualitative data was more revealing about the students' impressions about assigning competency. The following quotes are from various students addressing why I should continue to assign competencies:

> They are amazing. If you're behind in homework log it can help you earn a few extra points. I can have a specific goal to shoot for at the end of the day... You know what your goal is for that day. It's like you feel so good about yourself when you get a competency, it helps you want to go to math class the next day. It's something that you can work for which is within your reach. They help the students that are struggling and it gives you a self-esteem about math that White students already came into the class with! (Ninth grade African American girl)
>
> ★★★
>
> They motivate people and spark a learning spirit in the more lazy, laid-back people. (Ninth grade Asian American boy)
>
> ★★★
>
> If someone did really good one day, and you actually saw them challenging themselves more than others, I think they deserve more credit. (Eleventh grade African American boy)
>
> ★★★
>
> It's a good way to remind the class of the goal of the day and also some people who try really, really hard still don't get great grades so this gives them extra credit for what they *are* good at. (Ninth grade White girl)

It is clear from these quotes that self-esteem and feeling good about oneself are important aspects of learning mathematics, and the students enjoyed having their mathematical strengths publicly recognized. While assigning competencies may not have addressed my original goal of creating more equitable status among students, as is illustrated by the comment from the Asian American student about "lazy... people," their responses to the questionnaire illustrated that students felt better about their own achievement and felt more self-confident after I began assigning competencies. One possible explanation for the fact that students' views of who was good at mathematics did not change may be that I did not make explicit the connection between the idea of being good at mathematics and being good at the competencies. Since students tended to differentiate between being competent at something mathematical and getting good grades, perhaps students would have to see a long-term shift in grades in order to value the competencies.

Finally, a comparison of third-quarter grades (during which I assigned competencies daily) to second-quarter grades indicated that 13 students had improved their grades. Eleven of the 13 were African American or Latino; seven students improved their grades by one letter grade or more. Therefore, using competencies, possibly in combination with other interventions, had a positive impact on the performance of the African American students, which was the goal of my project. While they may not have gained "good at math" status in the way I had hoped, they had definitely gained tacit status by raising their level of achievement in the class.

Looking Back

What I learned about equitable teaching practices through my action research project informs my current teaching practice. My exploration of CI has had a significant impact on the way I teach. I do not officially assign "competency" any more, despite the fact that it showed benefits especially for African American students with lower status. It is not a practical approach to take in all of my classes. However, my Project IMPACT work raised my consciousness about status and the *teacher's* role in constructing competence.

As a result, assigning competence to students who have lower status has become informally integrated into my teaching practices. For example, one of the students in my ninth grade level mathematics course, an African American twelfth grader, had low status in the class and received an F for the semester due to poor classroom participation and performance. I have come to recognize his strength in quick mental calculations and estimation, and now constantly praise this student publicly for his competence.

In the second year of Project IMPACT, I experimented with CI's strategy of group roles to reorganize power structures among members of a group. More often than not I would see the same students (mostly White high-achieving students) step up and facilitate their group as they solved a problem, often leaving others behind or not making sure that everyone's questions were answered. Often those left behind were African American, Latino, or lower-achieving students. I found that the use of group roles helped many students learn more and yet for others, often those who were already high achieving, it had little effect. What was particularly interesting about the introduction of roles into my classroom structure was that some of the students started to complain about them. As a result, I decided to hold a class discussion about why I had implemented them. Some of the students said that I hadn't been clear about why I was changing things, that they worked really well as groups, and that this process wasn't needed. One of the students (a middle-class biracial [White and Mexican] high-achieving student who identifies as White) said,

> Are you doing this because the White kids are the main ones who pretty much lead their groups? I know I do and that's what I see other White students doing. If that's why, then I think it's a good idea, but you need to be clearer with us!

In response, I decided to have a discussion with the class about how the role of facilitator (for example) isn't something that you're born with, but rather is something you're implicitly or explicitly taught. Interestingly, the discussion seemed to help many of the students realize why I was changing the structure of our class and using roles. I think constantly keeping the idea of equity and increasing everyone's opportunity to learn on the table would have improved the use of CI in the classroom because most of my students cared about being fair and helping everyone learn.

Despite my commitment to teaching mathematics for social justice, these action research projects would not have taken place without a structured, inquiry-based, facilitated process with colleagues committed to the same goals. I believe many teachers share the same vision of reaching students who are being marginalized in our schools, but often they do not have the resources or support to take the first step toward changing their classroom practices. Students necessarily play an important role in classroom change as well. My experience has been that many students want an equitable classroom and that those in positions of power are generally willing to step back to allow other students to play a more prominent role.

Discussion

These studies speak to the promise of status management strategies like CI, as both a focus of professional development for equitable mathematics instruction and a pedagogical tool to restructure opportunities to learn in mathematics classrooms. In the first study, CI strategies helped shed light on how patterns in students' participation depended heavily on the way the classroom was organized. As youths whom teachers viewed as "less capable" than their peers were explicitly validated for making important contributions to their groups, they began to contribute more of their ideas to group and class discussions. This, in turn, shifted teachers' perspectives on what these students—marked by particular institutional, cultural, and racial labels—could do in their classrooms. In other words, CI created opportunities for teachers to envision and bring about greater relational equity among their students.

The shift in focus from what students are *not* doing to what they are *capable of* doing is a fruitful step toward exposing and challenging entrenched perspectives of children from nondominant backgrounds as inherently deficient in the skills and dispositions necessary for school success. We argue that the "deficit" perspective potentially blinds teachers to the impact of systemic injustice on students' opportunities to learn. We've also noticed that although teachers may hold the belief that the current educational system is broken and inherently inequitable, and desire to redress this injustice in their own classrooms, they may not apply this logic to the level of individual students. In other words, students who do poorly in tests, rarely volunteer to answer questions, sit back during group work, or even act out in oppositional ways may still be viewed solely in terms of individual ability and motivation, instead of as situated within broader social, cultural, and educational systems that necessarily shape their participation.

This chapter also illustrates how changes in students' participation can be *sustained* over time through practices such as assigning competence and group roles that have the potential for challenging stereotypes of groups of students. In the second study, Quindel brought an understanding of the relation between classroom participation and status, hierarchical structures of power in society and issues of equity to her inquiry project. Hence, instead of grappling with the participation practices of individual students, her efforts were geared toward disrupting unequal power relationships that she perceived were being remade in her mathematics classroom. Her experimentation over time with various status-managing techniques upset typical patterns of classroom participation between students from nondominant ethnic and racial groups and White students. While the students themselves did not necessarily change their views on the "smartest" students in the class, they did recognize that leveraging students' strengths and making each student integral to the work of the group helped them feel more motivated and confident. They no longer viewed their classmates in terms of one-dimensional characteristics (Horn, 2007), and even acknowledged the importance of reorganizing opportunities for group leadership among different racial groups.

There are additional aspects of Quindel's study that we find worthy of further discussion. One is the apparent disconnect in students' perceptions between being good at mathematics and being good at the competencies that comprise mathematical activity. Unlike the students in a study by Boaler and Staples (2008), who, over the course of four years, came to describe mathematical competency in terms of classroom practices such as justifying, explaining, and assisting others, students in this study may not have had the time to begin to question the role of grades as the traditional and sole marker of mathematical achievement. A second point is that the tension that arose between students' perspectives on group roles allowed the teacher to bring the relation between structure and agency to the students' attention, by talking about how one could learn to be good at roles given the right opportunity and practice. What we take away from both of these results is that it may be important for teachers to help students explicitly grapple with entrenched notions about ability and competence that reflect what they have come to expect from their prior schooling experiences.

We also want to point out that while both studies drew on CI methods (primarily because of the authors' familiarity with them), we recognize that other interventions could have been successful as well. CI is a set of fairly straightforward pedagogical structures that when understood by teachers can be seamlessly integrated into their classroom. This is not to say that developing group-worthy tasks is a trivial matter, nor that assigning competence arises naturally as teachers work with groups around these tasks. However, these pedagogical structures are aligned with constructivist theories of mathematics learning—the underpinning of current reform efforts—and as such do not require a significant shift in mind-set or curriculum for reform-minded teachers. Thus, it is less important to adhere to a particular pedagogical method—such as CI—than it is to understand the rationale behind this method and to develop a set of strategies that are driven by this rationale. For example, as Quindel notes in her reflections, techniques such as assigning competence served primarily to reaffirm the importance of issues of status and marginalization in her classroom. While she no longer employs this particular method, she has embedded her own techniques for acknowledging multiple competencies into the flow of her teaching.

At the same time, the use initially of a well-organized, research-driven pedagogical method for professional development purposes aided teachers in framing and talking about their work with each other. CI provided a set of ideas and language that they could draw upon to place and describe aspects of their classrooms that they thought might be relevant to issues of equity. This shared discourse also helped them to find ways to talk about things differently.

We believe it is also significant that both professional development efforts were based on a "working group model," in which teachers worked alongside other teachers, teacher educators, researchers, and graduate students to design and carry out interventions in their own classrooms. Professional development programs often fail because teachers are not necessarily invested in overall program goals, nor the changes required to implement the program faithfully in their classrooms (Ball & Cohen, 1999; Putnam & Borko, 2000). In contrast, the working group model is characterized by group negotiation, reflection, and iteration, whereupon teachers necessarily take the lead in deciding which approaches are most effective for their individual classrooms. Inserting teachers into these roles is a powerful way of *rearranging status* within the context of professional development, and education more broadly.

A final point we want to make about these studies is that CI and status-management approaches like it can be perceived as colorblind and colormute (Pollock, 2004) approaches to equity. We realize that their use may serve to mask rather than reveal issues of racism, White privilege, and power in education. The second study illustrates how these strategies can be taken up in color-conscious ways, but it is up to individual teachers to bring this lens to their practice. Clearly, additional, longitudinal research is required to assess whether status-managing techniques can lead teachers to question why students from nondominant groups are marginalized in the first place. What these two studies serve to do, however, is to enhance our understanding of how the disruption of traditional status hierarchies, in both our classrooms and professional development work, can give rise to legitimate and meaningful participation of marginalized groups of individuals toward greater human potential.

Note

1. Pseudonyms are used to refer to all participants.

CHAPTER 10

Creating "Constructive Opportunities": A "How" to Embracing Students' Mathematical Conceptions

VANESSA R. PITTS BANNISTER,
GINA J. MARIANO, AND CARLA D. HALL

During the past decades, mathematics education research related to "knowledge for teaching" has undergone virtual revolutions in response to nationally developed standards (e.g., NCTM, 1989, 1991, 2000) and reports (NCEE, 1983; NRC, 1989, 1990) that greatly expand expectations for both content knowledge and pedagogy. As a whole, these standards and documents call for a curriculum that prepares all students to be mathematically literate (i.e., to use mathematics to communicate, problem solve, and reason) and instructional practices that embody student-teacher interactions, higher-order cognitive thinking, problem solving, and discovery learning. In recognizing the disparity in mathematics achievement between minority students and their White counterparts (Braddock & Dawkins, 1993; Secada, 1992; Tate, 1997), National Council of Teachers of Mathematics (NCTM) (2000) emphasized the issue of equity and introduced the Equity Principle into its set of principles for mathematics teaching and learning: "Excellence in mathematics education requires equity—high expectations and strong support for all students" (NCTM, 2000, p. 12). By introducing the Equity Principle, NCTM (2000) brought to the forefront an important dimension of "knowledge for teaching"—high expectations. (See Foster, 1986; Hilliard, 1989; Jamar & Pitts, 2005; Knapp, 1995; Oakes, 1990; Solomon, Battistich, & Hom, 1996 for further discussions of teacher expectations.) Accordingly, the nature of past mathematics curriculum and instruction of minority students as focusing on "basics" to the exclusion of important aspects of mathematics learning (Oakes, 1985, 2005) is unacceptable. In particular, merely basic skill proficiency is not enough for "true knowledge and mastery of mathematics" (Secada, 1992, p. 630). Moreover, those strategies that are antithetical to constructivist principles—in particular more teacher-directed instruction and less student-led exploration; little cooperative and peer-supported learning; and more structured, lecture-style presentations (Knapp, 1995)—are considered appropriate when working with minority students.

As mathematics teacher educators, we face difficult questions about how to prepare teachers to understand and actualize the essence of high expectations. In particular, how do we help teachers to manifest facets of high expectations in their work and deeds as advocated by NCTM (2000)? For instance, how do we begin to assist teachers in their efforts to embody the following elements of high expectations (Jamar & Pitts, 2005): (a) to use students' prior knowledge as building blocks to new knowledge—which lets students know that they already have the foundation needed to learn, (b) to expect students to be active participants in their own learning—which lets students know that they are responsible for their own learning, and (c) to provide opportunities for students to *understand* concepts prior to learning rules—which lets students know they *can* understand the content and that it is understandable.

In this chapter, we argue that in order for teachers to have high expectations for their students, professional development should provide teachers with fruitful opportunities to revisit mathematics topics and learn effective ways to embrace students' conceptions. We will discuss explorations from professional development sessions that aimed to advocate (a) developing a deeper understanding of rational numbers and presenting lessons in richer ways tied to that understanding, and (b) embracing diverse mathematical conceptions held by students. In the following sections, we first describe the professional development sessions. Next we provide detailed examples from the sessions. And we conclude with a discussion of the interdependent facets of the sessions and the important role that such sessions play in the efforts to help teachers acquire appropriate understandings of present-day propositions of mathematics learning and teaching.

Description of Professional Development Sessions

The professional development sessions discussed in this chapter involved a university mentor; an urban middle school mathematics teacher, Ms. Alexander;[1] and two middle school students. The sessions were a subset of a year-long professional development effort. The initial professional development sessions involved discussions that focused on mathematical tasks, involving rational numbers, completed by the teacher. The teacher and university mentor met daily for about two months. Each session focused on increasing understanding of rational numbers and solution methods presented by the teacher. Separate subsequent sessions involved discussions that focused on mathematical exercises completed by the two students. Sessions with students took place twice a week for about one month.

The procedures utilized in the sessions with the teacher and students were similar to that of process tracing (Henderson & Tallman, 1998; Maryland, Patching, & Putt, 1992). Process tracing has been cited frequently in decision-making literature (Williamson, Ranyard, & Cuthbert, 2000; Covey & Lovie, 1998; Biggs, Rosman, & Sergenian, 1993) and is often used by researchers to identify reasoning characteristics and strategies (Patrick & James, 2004). Process tracing encompasses a wide variety of techniques that aim to reveal how an incident unfolds and how participants interpret the incident (Woods, 1993). The goals within the professional development sessions followed these principles in that the researchers sought to understand how the teacher and the students "unpacked" mathematics problems and interpreted and understood these problems. Several process tracing techniques were modified and used within the professional development sessions. For instance, process tracing usually involves collecting verbal reports from the task performer during or after performance (Patrick & James, 2004). In a similar manner, we used several examples of verbal reports, including modified versions of stimulated recall and "think aloud." Stimulated recall elicits learner commentaries retrospectively with the support of visual/audio stimuli and has been used to gain insight into learners' cognitive processes (Egi, 2008). Our approach was different with respect to individuality because the focus was to provide the teacher and the students with opportunities to explore their own understandings of mathematical concepts; therefore, interactions during the sessions were based on the individual(s). Some sessions involved the teacher or the student explaining his/her thought processes while solving problems, while other sessions involved problems being solved initially, followed by explanations. The structure of the sessions was dependent on the individual and his/her needs and understandings.

With respect to "think aloud" protocols, in which students are instructed to verbalize their thoughts while completing a task, or answer questions in a stimulated recall interview (Mao & Benbasat, 1998; Butcher & Scofield, 1984), we asked the teacher and the students questions that encouraged them to explain their methods of solution. This method is similar to that of Wang, Hawk, and Tenopir (2000). Wang and colleagues (2000) used computer screenshots concurrently with voice recordings while graduate students answered questions using the Internet as a reference in order to understand the cognitive processes students use when searching the Internet. Like Wang and colleagues (2000), we aimed to have the teacher and the students verbalize their thoughts while solving problems as it allowed session participants to capture instances of understanding and misunderstandings.

The strategies used during the professional development sessions involved a set of tools for the exploration of conceptions and/or misconceptions of mathematics concepts and ideas. Due to the nature of the sessions, which provide constructive opportunities for participants to openly share mathematical conceptions and garner deeper mathematical understandings via evaluations of individual and/or other individuals' work, we refer to them as *Constructive Opportunities*.

Constructive Opportunities with Teachers

In the following passage, Mr. Sherman, another teacher who participated in the professional development, describes qualities of teaching that contribute to an equitable mathematics classroom:

> You'd have a teacher who also listens carefully to what kids are saying, sort of both, sort of verbal and nonverbal kind of communication. A teacher who's presenting curriculum in a coherent and in a fashion that kids can comprehend. Comprehension? Someone who's setting the bar a little higher than where the kids are now. So the kids have to reach to get over it. But it's not so high, that they look up and say, "There's no way that I can do that." A place where there's support for kids, again that can be from the teacher, from a tutor, from, from classmates. A place

that, where there's some celebration. Celebrating when we're learning something.

Mr. Sherman suggests that equitable classrooms are places where (a) students' concerns and conceptions are valued, (b) teachers comprehensibly teach content, (c) teachers have high yet reasonable expectations, and (d) learning is celebrated. In this section, we propose that as teachers have multiple opportunities to revisit mathematics topics, they encounter a plethora of chances to examine ways to adopt the practices outlined by Mr. Sherman. In addition, they will reexamine aspects of teaching such as comprehension, transformation, instruction, evaluation, and reflection (Shulman, 1987). This re-examination may include activities highlighting exploration of comprehension, encouragement of continual transformation of understandings, acceptance of new knowledge to be reflected in instruction, and continual evaluation and reflection on current understandings to develop deeper and richer understandings. At the core of these activities is comprehension, for "to teach is first to understand" (Shulman, 1987, p. 14). Moreover, teachers' knowledge plays a vital role in what is done in classrooms, and most importantly, what students learn (Fennema & Franke, 1992; Heaton, 1992; Mullens, Murnane, & Willett, 1996; NCTM, 1991, 2000; Putnam, 1992).

Aspects of Constructive Opportunities with Teachers

Constructive Opportunities with teachers are based on trust. Within these situations, teachers and mentors should be comfortable with exploring mathematical topics in ways that advocate developing a deeper understanding of the underlying mathematics and presenting lessons in richer ways, tied to that understanding. In some cases, explorations may involve activities that are perceived as "easy," yet require a teacher to view particular mathematical ideas in varied ways. For instance, the exercise in Figure 10.1, which may be perceived as straightforward, helped create a constructive opportunity for Ms. Alexander to revisit ideas related to fractions.

Her initial comment about this task was as follows:

> "On this problem I can see this shape here (pointing to the second figure within the first figure). As I said, I know it's going to be...six equal parts of seven (takes a pause). So, actually, I got my eye on a nice chunk out of the whole (pointing to a portion of the first figure). I see it. It's close by (takes a pause). Can't see it!

172   *Mathematics Teaching and Learning in K-12*

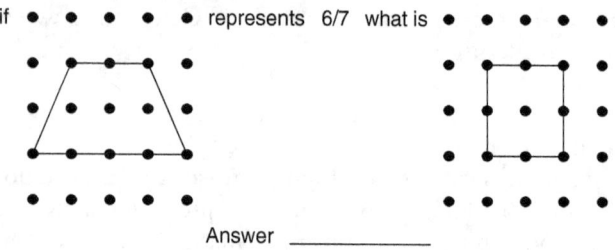

Figure 10.1 Exercise from dot activity.

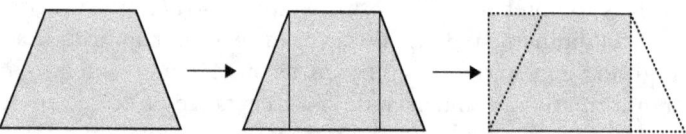

Figure 10.2 Ms. Alexander's first solution method.

After a discussion of similar problems, Ms. Alexander and her mentor[2] revisited the problem and had the following discussion about her method of solution. She began by explaining the first representation of her solution method as seen in Figure 10.2.

> *Ms. Alexander (A):* I transferred this over to here.... I wanted to do that so it would be easier for me to see.
> *Mentor (M):* Oh, that's good. So, what does the figure on the right side represent (pointing to the white portion of the third figure)?
> *Ms. A:* Over here, it represents half of this (referring to the left portion of the second figure).
> *M:* Half of what?
> *Ms. A:* I have one whole of this (pointing to the first figure), you know this piece here (pointing to right side portion of the second figure), could be here (pointing to transposed portion in the third figure), with this other section here is just half of that. Does that make sense to you?
> *M:* OK, so what does that represent (pointing to the third figure)? What fraction?

At this point, Ms. Alexander began to create equal parts of her representation (see figure 10.3).

> *Ms. A:* Well, I'm looking. Now I have one, three and this is six-sevenths (counting the squares in figure 10.3).

Creating "Constructive Opportunities"

Figure 10.3 Ms. Alexander's second solution method.

M: What's six-sevenths? Count six-sevenths? The boxes?
Ms. A: I had to do that in order so I could see this.
M: That's right.
Ms. A: That why I was looking to see—
M: So, how many boxes? OK, go ahead.
Ms. A: Over here, I have, one, two, three, four, five, six.
M: Mhmm
Ms. A: Six of these boxes, so then I'm looking at that, so—
M: Tell me what you're thinking.
Ms. A: Soon as I figure it out. Soon as I figure it out (laughing).
M: Pardon me (laughing). OK, let me see what you wrote there. Six-sevenths times one-sixth. Why did you do that?
Ms. A: Because I had six equal parts here, so I took it and took six-sevenths and figured out what each of these little pieces would be, and it would be one-seventh. That would finish the calculations and so each one of these pieces would be one-seventh and left four of these over here and that would be four-sevenths. OK?
M: Very good.
Ms. A: So, I can see that! Who thinks anymore in math?
M: No?
Ms. A: I'm just saying, I'm so accustomed, at least I am, to doing straight computation.
M: Mhmm.
Ms. A: And when I have to think it out, because it's always right there, and I guess I teach in that fashion. So this is foreign enough that I need retraining in it, so that I can see it. And that way, when I present it, I'm strong enough to say this is the part, uh huh, OK.

Throughout the sessions, the mentor exhibited key facets of *Constructive Opportunities* by (a) showing support for (or embracing) Ms. Alexander's responses (e.g., "Oh that's good. So, what does the figure on the right represent?"); (b) expressing interest (or caring) about Ms. Alexander's thinking (e.g., "Six-sevenths times one-sixth. Why did you do that?"); and (c) providing guidance as needed (e.g., "What's six-sevenths? Count six-sevenths? The boxes?").

Another important aspect of the sessions is when Ms. Alexander brings to light issues related to mathematics teaching and learning. She begins by proclaiming, "Who thinks anymore in math?" Her subsequent comment, "I'm just saying, I'm accustomed, at least I am, to doing straight computation," reveals she has experienced mathematics learning and teaching in ways that highlight procedural knowledge. (See a detailed discussion of procedural knowledge in Hiebert & Carpenter, 1992.) Although her experience was different from present-day propositions of mathematics learning and teaching (e.g., NCTM, 2000), Ms. Alexander appreciates that mathematics learning and teaching require thinking and are more than "straight computations." As she declares, "So this is foreign enough that I need retraining in it, so that I can see it. And that way, when I present it, I'm strong enough to say this is the part."

Ms. Alexander's discoveries create a foundation on which to build new knowledge structures aimed to acquire a "threshold level of subject matter knowledge" (Darling-Hammond, 1997, p. 308). Such knowledge will support her authentic efforts to effectively teach students and embrace their thinking (i.e., take account of students' ways of thinking and understand the reasoning behind students' thinking). The next section will discuss follow-up sessions aimed to assist Ms. Alexander in such efforts.

Constructive Opportunities with Students

To support and accommodate the mathematics learning of students, teachers must create opportunities for students to rediscover and reinvent concepts and methods of solution (Lappan, 1998). In view of this, teachers should consider

> engaging students in the task, pushing student thinking while the exploration is proceeding, helping students to make the mathematics more explicit during whole-class and group interaction and

synthesis, and using and responding to the diversity of the classroom to create an environment in which all students feel empowered to learn mathematics. (Lappan, 1998, p. 135)

Teachers should be perceived as "coordinators" of learning—not "main actors" (Lappan, 1997). Students should be encouraged to make sense of the stories they encounter in their everyday life, a reasonable parallel for understanding the goals of the *Standards*. As students engage in these roles, teachers should embrace students' ideas and appreciate their attempts as mathematics learners.

Aspects of Constructive Opportunities with Students

Constructive Opportunities with students should be organized to focus on student understanding. Within these professional development sessions, participants should consider students' work on particular tasks and ponder the patterns of solution methods, and what those patterns suggest about what students know. Constructive opportunities are not organized to address students' conceptions, but rather to gather data to analyze them. Initially teachers may pair with mentors to complete the sessions. Mentors may help teachers to (a) assess student work (before each session), (b) outline appropriate questions to ask students (before each session), and (c) collect data via video or note taking. In the following sections, we will discuss two sessions.[3]

A Constructive Opportunity with Willie. Willie was a student who could be perceived as an attention seeker. For instance, during classroom discussions, he was often very engaging and willing to offer answers and/or explanations that highlighted his misguided notions of concepts or ideas. He continuously stayed on task during group activities and completed (yet often incorrectly) individual work during class. As Ms. Alexander reviewed students' work on a particular task, she was intrigued by Willie's responses. In an interest to discover *how* Willie approached the problem, she asked him to be a part of a *Constructive Opportunity* session with her and her mentor. Willie happily accepted the invitation. What follows is an excerpt of the discussion between Ms. Alexander and Willie (W) as he explained how he determined the smallest and largest numbers from among $4/7$, $5/6$, $9/14$, and $2/3$.

W: I took each of them and got in my head what they look like and the fractions and I was like, hey, I see that $9/14$ you get most of the pie, I mean less of the pie then you have $2/3$ 'cuz you'd be getting most of the pie and $5/6$ you'd only leave one left, but $4/7$ you'd leave three left. But with $9/14$ you only leave, ahhh, what was it?

Ms. A: Five, is that what you're thinking of?

W: Yeah five.

Ms. A: What about $2/3$?

W: The $2/3$, that would be—

Ms. A: Would that not be one piece, like $5/6$?

W: Yeah. Those two would only leave two pieces. I mean, one piece and $4/7$ would leave three, I was thinking, hey, this is the smallest, you'd leave, you'd leave the most out of it if you were to eat certain pieces, that's how I got my answer. I couldn't explain it at the time, it was kinda hard. I thought it over in my head, some kind of method.

Ms. A: How did you determine which one was larger between $5/6$ and $2/3$, because there's still one piece left from each one, right?

W: Mhmm.

Ms. A: So, how did you determine $5/6$ was the largest out of the four fractions there?

W: I kinda, I saw that $5/6$ and $2/3$ were exactly the same, so I was thinking at the time it doesn't matter, they're pretty much the same. I wasn't too worried about which one would be smaller out of the two, I was thinking they being the same, obviously. So, I put one out of the two.

(brief discussion with another student)

W: I did it in my head.

Ms. A: OK, that's good.

W: Something I learned in elementary. My teacher taught me, which would you rather have, $3/4$ of the pie or $9/16$? And I was like $3/4$ because you get more of the pie, because there's not that much left. If you get $9/16$ and everybody get[s] the rest, what you [have] left is not a lot. Either way it's a lot, but I'd rather have two out of three, there's more for me.

At first read, this may be perceived as a missed opportunity to address misconceptions. Instead, this should be considered as an opportunity to deeply consider Willie's work on the task and ponder his solution

Creating "Constructive Opportunities" 177

methods, and what those patterns suggest about "what he knows." Ms. Alexander exemplified this level of concentration as she attended to Willie's ideas regarding the task and mathematics at hand. Ms. Alexander embraced Willie's ideas by (a) attentively listening (e.g., "Five, is that what you're thinking of?"); (b) tracking his ideas (e.g., "What about $2/3$?" "Would that not be one piece, like $5/6$?"); and (c) posing questions that create opportunities for discussions (e.g., "How did you determine which one was larger between $5/6$ and $2/3$, because there's still one piece left from each one, right?").

A Constructive Opportunity with Katie. Katie was a very shy student. She rarely participated during group activities or completed individual work during class. This kind of student may be perceived as lazy or unmotivated. From an interest in understanding *how* Katie approached the task that she discussed with Willie, Ms. Alexander asked Katie to be a part of a *Constructive Opportunity* session with her and her mentor. Katie, unlike Willie, was hesitant to accept the invitation, but eventually agreed to participate. Katie provided the answers to the problem as seen in Figure 10.4.

When asked to explain the responses to the work shown in figure 10.4, Katie provided the following illustrations of the fractions as seen in figure 10.5.

Katie (K) presented two explanations as to why $2/3$ and $9/14$ were the smallest and largest numbers, respectively. Using Figure 10.4, she presented the argument that since 2 (the numerator of $2/3$) is less than the other numerators (4,5 and 9) and 3 (the denominator of $2/3$) is less than the other denominators (6, 7 and 14), $2/3$ is the smallest. Similarly, since 9 (the numerator of $9/14$) is greater than the other numerators (2, 4, and 5) and 14 (the denominator of $9/14$) is greater than the other denominators (3, 6, and 7), $9/14$ is the largest. She continued to present this line of thinking as she explained her illustrations presented in figure 10.5.

> *Ms. A:* I have one more question for you. This (pointing to Katie's illustration of $2/3$) represents $2/3$, can you tell me why this (pointing to Katie's illustration of $2/3$ again) is a smaller box than this one (pointing to Katie's illustration of $5/6$), that one (pointing to Katie's illustration of $9/14$), and this one (pointing to Katie's illustration of $4/7$)?
>
> *K:* Because this one (pointing to her illustration of $2/3$) is three, three portions and I just shaded two. And this one (pointing to her illustration of $5/6$) is six portions and I shaded five.

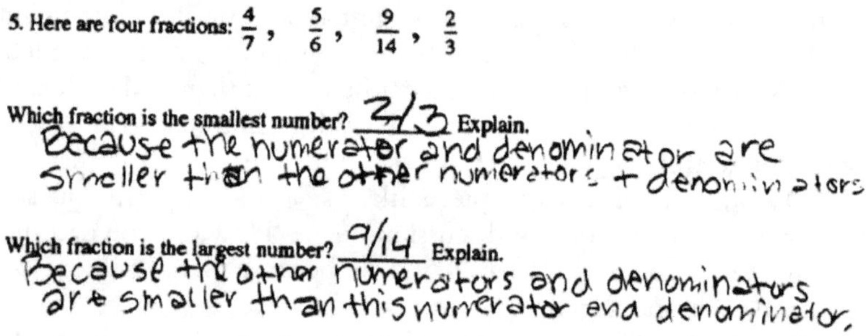

5. Here are four fractions: $\frac{4}{7}$, $\frac{5}{6}$, $\frac{9}{14}$, $\frac{2}{3}$

Which fraction is the smallest number? __2/3__ Explain.
Because the numerator and denominator are smaller than the other numerators + denominators

Which fraction is the largest number? __9/14__ Explain.
Because the other numerators and denominators are smaller than this numerator and denominator.

Figure 10.4 Katie's responses.

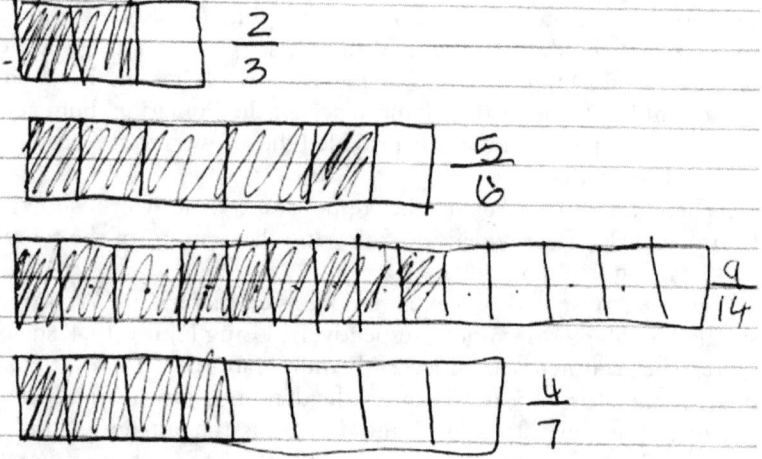

Figure 10.5 Katie's fraction illustrations.

Although Katie did not provide as many verbal responses as Willie, Ms. Alexander continued to embrace Katie's methods of explaining her solutions. Specifically, Ms. Alexander supported Katie's use of illustrations. Consequently, opportunities were created for Katie to openly share her ideas (in her own way) and for Ms. Alexander to gain a deeper understanding of Katie's thinking.

Facets of Constructive Opportunities

In this chapter, aspects of *Constructive Opportunities* with respect to participation between teacher and mentor (i.e., sessions with

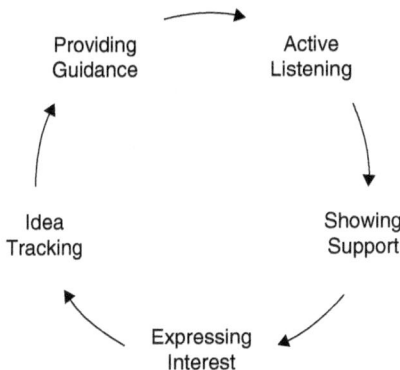

Figure 10.6 Core facets of *Constructive Opportunities*.

Ms. Alexander and Mentor) and teacher and student (i.e., sessions with Ms. Alexander and her students—Willie and Katie) were described. The aspects depicted include actively listening, showing support, expressing interest, tracking ideas, and providing guidance (see Figure 10.6).

Although the facets were described separately for students and teachers, many interact interdependently in order to create an environment that embraces individuality with respect to mathematics thinking. Moreover, the relationships between these facets have reciprocity. For instance, actively listening to teachers' and/or students' explanations while solving problems, showing support for their ideas, and expressing interest regarding their thinking allows mentors and teachers to track ideas and thought patterns in order to provide appropriate guidance. Accordingly, in *Constructive Opportunities* teachers, mentors, and students participate in a continuous cycle in which the facets remain the same, but the features of each step change on the basis of the characteristics of the participants.

Conclusions

The implementation of strategies such as *Constructive Opportunities* geared toward the new vision of teaching and learning as suggested by nationally developed standards (e.g., NCTM, 1989, 1991, 2000) and reports (NCEE, 1983; NRC, 1989, 1990) is difficult and should not be expected to be otherwise. Findings from explorations involving facets of *Constructive Opportunities* provide a window of opportunity to begin conversations between mathematics educators about effective strategies to understand, address, and broaden mathematical conceptions of teachers and students. Moreover, engagement in *Constructive Opportunities* could assist teachers (as it did Ms. Alexander) to acquire productive methods to embrace students' mathematical conceptions, thus manifesting elements of high expectations as suggested by NCTM (2000) and Jamar and Pitts (2005). To ignore the role of professional development efforts such as *Constructive Opportunities* in achieving the goals of the Equity Principle (NCTM, 2000) "is to be like the man who was looking for a lost coin two blocks away from where it was lost because the light was better at the new spot. If he were to continue in that way, the problem would never be solved" (Hilliard, 1974, p.44).

Notes

1. The names of all participants, both teachers and students, are pseudonyms. All participants were involved during the course of the data gathering with the Diversity in Mathematics Education (DiME) Project at the University of California, Berkeley.
2. The mentor is the first author of this chapter and participant of the Diversity in Mathematics Education (DiME) Project at the University of California, Berkeley.
3. The first author (the mentor) assisted Ms. Alexander with the two sessions during the Diversity in Mathematics Education (DiME) Project at the University of California, Berkeley.

CHAPTER 11

Using Lesson Study as a Means to Support Teachers in Learning to Teach Mathematics for Social Justice

TONYA GAU BARTELL

A growing number of researchers argue that teaching mathematics for social justice can support the ongoing struggle for equity in mathematics education (Frankenstein, 1995; Gutstein, 2006). One way to support mathematics teachers in developing mathematics pedagogies for social justice may be to use lesson study, a powerful form of professional development in Japan that is increasingly used in the United States (Lewis & Tsuchida, 1998), to engage them in goal-oriented, reflective practice. To date, no published studies exist suggesting lesson study has been used in the context of supporting teachers in developing pedagogies for social justice. Moreover, little research exists that examines mathematics teachers learning to teach for social justice. This chapter reports on a study of eight secondary mathematics teachers who participated in a graduate course that engaged them in a version of lesson study to create, implement, observe, revise, and reteach math lessons that incorporated social justice goals. The research reported examines the challenges teachers faced in learning to teach mathematics for social justice and the ways the use of lesson study may have supported or constrained teachers' negotiations of these challenges.

Teaching Mathematics for Social Justice

What social justice teaching "actually means is struggled over, in the same way that concepts such as democracy are subject to different senses by different groups with sometimes radically different ideological and educational agendas" (Apple, 1995). Varying goals, content, and audiences affect one's conception of socially just teaching (Chubbuck & Zembylas, in press), and different conceptions might actually limit approaches toward justice and equity (North, 2008).

The conceptualization that informed this work draws especially from the work of Freire (1993), who conceived of a critical pedagogy which would support humans in actively transforming their society to make it better. Drawing on Freire and, in the context of mathematics education, on the work of Gutstein (2006), I conceptualize teaching mathematics for social justice as based on three fundamental ideas. First, mathematics can be used to teach and learn about issues of social injustice or to *read the world* (Gutstein, 2006). Mathematics is used to develop critical consciousness and to support students in deepening their knowledge of the sociopolitical contexts of their lives. Second, mathematics is used in this way to ultimately support action toward equity, or to *write the world* (Gutstein, 2006). Mathematics is used to not only understand the world, but also to change it. Finally, through the process of using mathematics to analyze and change their world, students strengthen and extend their knowledge of mathematics.

Lesson Study & Learning to Teach for Social Justice

One way to support mathematics teachers in learning to teach for social justice may be to use lesson study to engage them in goal-oriented, reflective practice. The fundamental components of lesson study include teachers formulating long-term goals for student learning, collaboratively developing and implementing lessons toward these goals, observing the implementations, and revising the lessons as necessary with respect to the achievement of these goals (Stigler & Hiebert, 1999). This section outlines these components and begins to connect lesson study to research around teachers learning to teach for social justice.

Identification of Shared Goals

Lesson study is an inherently collaborative process that begins with teachers identifying communal goals for student learning. In the context of learning to teach mathematics for social justice, teachers must agree not only on the mathematical content goals, but also on the social justice issue of focus. A primary component of teachers learning to teach for social justice is developing an understanding of the social contexts influencing teaching and learning (Cochran-Smith, 1999; Darling-Hammond, 2002). Teaching and learning are linked to economic, political, and social power structures in society. Affected by these power relations, schools and classrooms are not neutral sites. Teachers learning to teach for social justice grapple with comprehending how these structures interact with their understanding of teaching and learning (Cochran-Smith, 1999) and, as teachers engage in the lesson study process and think about the mathematical and social justice goals for their lessons, they will also likely grapple with examining how these power structures influence their decisions about which goals to pursue and whether these goals are met.

Access Multiple Resources

Once goals have been established, teachers engaged in lesson study work to jointly plan a small number of lessons aimed toward achieving those goals. Accessing multiple resources to improve one's own knowledge base is an important initial step in designing quality lessons. A primary resource is one's students. Knowing your students well includes understanding your students' current mathematical competencies in order to meet students where they are to help them learn. A chief component of lesson study that can sustain this focus is its careful attention to anticipating students' responses (Fernandez, Cannon & Chokshi, 2003). When designing lessons, teachers must consider what knowledge students are likely to bring to the task, what strategies students may use, and how students' knowledge connects to various mathematical concepts. Observations and reflections focused on student thinking then enable teachers to develop knowledge of how students think about and learn from the tasks they engage in, and to determine what makes certain learning experiences effective for students (Fernandez et al., 2003).

In the context of learning to teach for social justice, teachers must also know their students well in order to build on students' cultural and community knowledge and linguistic resources (Cochran-Smith, 1999). They must understand their students "in non-stereotypical ways while acknowledging and comprehending the ways in which culture and context influence their lives and learning" (Darling-Hammond, 2002, p. 209). Teaching for social justice involves making issues such as power, racism, and classism explicit parts of the classroom (Cochran-Smith, 1999). Students have a wide range of reactions to these issues, including excitement, anger, resistance, and immobilization (Griffin, 1997). Knowing your students includes anticipating the many responses students might have to such pedagogy.

Teacher Self-Reflection

Another fundamental component of lesson study is the teachers' self-reflection. "Rather than asking teachers to examine their practice with the premise that reflection leads to growth, lesson study asks teachers to plan, implement, and refine lessons with the premise that this leads to reflection" (Fernandez & Yoshida, 2000, p. 4). Teachers reflect on their practice in a variety of ways within this process, including incorporating feedback received in the debriefing sessions about their own teaching into their knowledge base or in contemplating their philosophy of teaching (Lewis, Perry, & Hurd, 2004).

As teachers learn to teach for social justice, they reflect on their understanding of themselves, both personally and in relation to others (Darling-Hammond, 2002). This includes reflecting on how their beliefs about teaching and learning are influenced by the cultural, historical, and economic contexts in which they grew up, and teachers trying to understand perspectives and experiences of others different from their own to reflect on how their personal biases may impact their teaching. As teachers engage in lesson study around teaching mathematics for social justice, they may see how other teachers make different assumptions about situations, that students do not respond to the lesson in anticipated ways, and their self-reflection may expand to include a reflection on how their own biases may affect the mathematics teaching and learning process.

This study reports what teachers communicate about the challenges that they faced in learning to teach mathematics for social justice and how the use of lesson study supported or constrained their negotiations of these challenges. The purpose of concentrating on the challenges teachers face is not to focus on ways in which teaching for social justice may be limited. Rather, examining challenges both reaffirms the complexity, time, and introspection required for teaching mathematics for social justice and underscores this as a necessary part of the learning process. In this effort to describe learning to teach for social justice, this research aims to go beyond simply describing what teaching mathematics for social justice looks like to examine how teachers comprehend it, make sense of it, and struggle with it.

Methodology

Data for this study were obtained from four secondary mathematics teachers (Holly, Roxy, Ann, and Dana[1]) enrolled in a 15-week graduate course on teaching mathematics for social justice. These four teachers taught at three area high schools, and all four teachers were white females. The teaching experience across these teachers ranged from 6 to 16 years,[2] and all four had a bachelor's degree in mathematics. During the same year as this study, all four teachers were engaged in a district-wide equity initiative in which they examined and discussed the racial achievement gap statistics for their school and in which they participated in mandatory sessions entitled "Courageous Conversations about Race" (Singleton & Linton, 2007), exploring concepts such as institutional racism and white privilege in the educational context. All four teachers described substantive involvement (e.g., attending meetings, leading discussions) in these equity initiatives at their school sites, and one of the teachers functioned as a representative for her school's equity team, facilitating discussions on equity.

Additionally, one of the teachers in this group had prior exposure to notions of teaching mathematics for social justice through involvement in a year-long professional development seminar the year prior to this study, and she expressed that she enrolled in this graduate course to engage explicitly with the ideas of math for social justice that had been introduced in this previous experience. Three of these four teachers were also enrolled in a Master's program at a local university where they engaged in discussions of equity in education (one of these teachers had also taken two doctoral-level courses on culturally relevant pedagogy), and all four teachers reported that they enrolled in the course because they were interested in thinking about teaching mathematics for social justice as a means to support their own and their school's efforts toward equity.

The Graduate Course

I taught the graduate course that served as the context for this study. The central activity for the first part of the course was discussion and analysis of readings focused on teaching for social justice in general and teaching mathematics for social justice specifically. Since all of the teachers were engaged in their district's various equity initiatives, the course did not focus explicitly on supporting teachers in examining themselves in relation to learning to teach math for social justice or in examining how the larger social, historical, and political dynamics of teaching intersect with teaching mathematics for social justice. Instead, this first stage focused on developing teachers' conceptions of what it might mean to teach mathematics for social justice, recognizing that teachers would still grapple with issues of self and social contexts as they engaged in these activities and discussions.

The central activity of the second part of the course was a lesson study, in which the teachers collaboratively designed a lesson reflecting their definitions of mathematics for social justice with consideration of anticipated student responses. Teachers taught these lessons and observed one another during these teaching sessions. After each lesson implementation, teachers met immediately for an approximately one-hour debriefing session to discuss the lesson's effectiveness in terms of student learning of the lesson's mathematical and social justice goals. Teachers then revised and re-taught the lessons.

Analysis

These analyses are guided by situated, sociocultural perspectives of teacher learning and professional development (Lave & Wenger, 1991). Situated, sociocultural theories of teacher learning center on the concept of learning as situated social practice, in which learning is an "emerging property of whole persons' legitimate peripheral participation in communities of practice" (Lave, 1991, p. 63). This shifts the focus to people engaged in mutual enterprise, with a shared repertoire of actions, discourses, and tools (Wenger, 1998). Teacher learning, then, is influenced not only by personal orientations, but also by teachers' interactions within various social communities.

Data were collected during a single semester and were analyzed systematically. The data analyzed for this chapter consist of (a) 12 meeting transcripts, (b) written teacher reflections, (c) lesson plan artifacts, (d) teachers' final course papers, and (e) pre- and postseminar teacher interviews for each teacher. I drew on grounded theory approaches (Glaser & Strauss, 1967; Strauss & Corbin, 1990), analyzing the data for recurring themes. In the analytic process, I made initial conjectures from the existing data record, and then continually revisited and revised those hypotheses in subsequent analyses. Data analysis involved a constant comparison of all these data in order to ensure that the larger claims made accurately reflected the evidence leading to those claims. The data were searched for both confirming and disconfirming evidence that might support or challenge a particular assertion (Erickson, 1986).

Results

The primary research questions enumerated here, what challenges teachers reported that they faced in learning to teach mathematics for social justice and whether and how lesson study supported or constrained teachers' negotiations of these challenges, guides this presentation of results. Challenges that arose in the context of identifying shared goals, accessing multiple resources, engaging in teacher self-reflection, and implementing and revising the study lessons are highlighted. In the course of presenting these findings, I highlight details about the lesson the teachers planned in the context of the lesson study process in an effort to provide readers with a sense of the groups' interactions and to put the results in perspective with respect to the issues that arose.

Identification of Goals: Early Tensions Arose

Early conversations during class time suggested a challenge in identifying the lesson's goals, particularly with respect to incorporating *both* mathematical and social justice goals in a single lesson. All four teachers commented early on about this tension during class sessions. Dana described the challenge as "figuring out how to teach for social justice while fitting in the required curriculum that meets state standards and testing." In addition, this tension was about determining what should come first when planning the lesson, the math topic or the social justice issue. As Ann said, "I guess the question is what comes first. You find a situation and you try to fit it to the curriculum or you look at the curriculum and you go we're going to be studying this." Roxy replied that for her, "if you're going to try to get to the SAT, the curriculum comes first." In her postinterview, Roxy again noted that a danger of starting with the social justice first is that "it may lend itself to a sort of artificial bending of mathematics... you're trying to fold the mathematics around some other content that you are trying to cover because you're trying to infuse social justice."

Regardless of which comes first, Holly and Ann expressed the importance of the mathematics. Ann felt that including relevant mathematics "makes it more palatable in terms of why are we doing this," and Holly felt that "kids would find it more valid if it included math." It was also important that this was rigorous mathematics. Holly, commenting about an example in a reading, said that what she liked about it was that "it was hard math. There was nothing soft about the mathematics." Later, Ann reminded her group, "I don't want the message about social justice to be social justice is for lower-level mathematics. It's what you do maybe in Algebra I, but then you go learn real mathematics and that's not useful for actually anything."

Accessing Multiple Resources: Prioritizing a Focus on the Social Justice Component

To deal with the challenge of negotiating both mathematics and social justice goals in a single lesson, these teachers tended to prioritize a focus on the social justice component during the lesson design process. To sustain this focus, they expressed a need to access resources in particular ways. To situate these results, I present necessary background information about the lesson itself.

Initial lesson design. Holly, Dana, Ann, and Roxy designed a lesson examining minimum wage versus living wage. The lesson design began with students completing a homework assignment, calculating basic measures of central tendency for data, and reasoning about which measure was most appropriate to use in different situations. The lesson, the next day, began by asking students to use local data to determine the average cost of housing for one person in their city. Students were then asked to complete some mathematical calculations to determine how much someone making minimum wage (working 40 hours per week) could afford to pay in monthly rent. Next, students were asked to calculate, given the average cost of housing for one person they had previously found, what hourly wage this person would need to make to afford that housing. In other words, teachers asked students to calculate the living wage. The subsequent planned class discussion centered on the discrepancy between minimum wage and living wage, asking students to brainstorm solutions that might alleviate the discrepancy. Finally, students were asked to complete one of three homework assignments for the next day. The first homework assignment asked students to write a "letter to the editor" in response to two recent letters that were published in the local paper; the second asked students to reflect on the day's lesson about what they had learned, about what they liked and disliked, and about any other aspects of the lesson they would like to comment on; and the third asked students to consider and respond to an excerpt regarding how the federal poverty line should be calculated.

During the initial lesson design phase, Holly, Dana, Ann, and Roxy articulated both mathematical and social justice goals for their students. In terms of mathematical goals, they saw this lesson as supporting students' review of basic statistics, ratio, and proportion. For example, Holly wrote in an early-semester reflection that the mathematical goals of the lesson are to "review mean, median, mode, range" and to "have a better understanding of proportion." The teachers also said that students' review of mathematical concepts in a new context to "read the world" (Gutstein, 2006) with mathematics was an additional goal. Holly remarked in class that in the teaching of the lesson her group had developed, she had thought, "students [would] step away from the lesson with a new outlook on how math can be an effective tool in their lives. It empowers them to solve critical, close-to-home problems."

In this initial stage, the teachers also identified social justice goals they had for their students with respect to this lesson, and these goals were rearticulated throughout the semester. First, teachers hoped that students would recognize that a minimum wage is not a living wage. Ann, for example, indicated that she wanted students to "know that there is a serious disparity between the wage required to live in [this city] and the wages paid to low-wage workers." She stated in class the following evening that "ultimately I don't care if they can come up with really what the average house costs.... The experience I want them to have is the number a minimum wage person can afford is nowhere near anything in here." Second, teachers wanted students to recognize that everyone should earn a living wage. Dana noted that the issue of the lesson is that "it's not right that somebody works 40 hours a week and can't afford to live." Third, the teachers wanted to make sure students did not interpret the lesson as a cautionary tale about individual betterment. As Ann stated in class, this lesson was not about "go better yourself so you don't have to have a minimum wage job." Roxy shared similar concerns, wanting students to "[look] at who does this affect" so that they could "take the perspective of other people."

A concentrated focus. As the teachers were planning their lesson, Holly proposed early on in a group meeting that it might be better to focus on the social justice component of the lesson. She remarked, "And really my last thought is the math. I think that if I could think of a topic that might be social justice-y, then we could think about the math that's in it later." Ann, in the same meeting, stated, "I don't necessarily need to tie a specific [math] objective in my course to it. That's going to be challenging.... I'm not saying that math has to drive it because, good luck."

At this same time, one of the central concerns for the teachers was a need to develop their own knowledge of the social issue of interest. Ann commented that her "knowledge of particular social issues is weak," and Holly wrote in a personal reflection after our second course meeting that one concern she had about teaching mathematics for social justice was "opening the door to the unknown because my knowledge on many issues that may arise is weak at best." Perhaps driven by this need to learn more, the teachers requested, and I provided, a list of websites to serve as a starting point for their investigation into social justice issues. These were websites for various organizations advocating teaching for social justice (e.g., www.radicalmath.org, www.rethinkingschools.org). The teachers also regularly brought in newspaper articles or other information that they found relevant to their lesson to share with the group. One teacher called a local organization to seek out information on Section 8 housing in the area, and another teacher contacted a local legislator for information about local housing issues. A third teacher suggested that the book *Nickel and Dimed: On (not) Getting by in America* (Ehrenreich, 2001) was particularly relevant to their discussion. She brought the book to multiple course meetings as a reference, and three of the four group members read the book, drawing from what they read to contribute to the lesson design. Large portions of the group members' meeting time during the first five weeks of the lesson planning process were spent discussing the information they read and refining the social justice goals for their lesson accordingly.

Teacher Self-Reflection: Getting Personal

Another challenge recognized by these teachers was how to wrestle with the effects addressing issues of social justice had on one personally. This challenge did not arise in any one moment of the course, but instead the teachers reflected continuously about things that happened in their own lives, past and present, and about how this course was influencing them personally. Roxy commented early on in the course during a class meeting that as she was reading she

> had this moment of despair where, how do you live the life examined and not just crawl into a hole? How do you really tackle teaching something like this and yet continue to live what is essentially a hypocritical life on a lot of levels?

Later that same evening, Roxy talked about the personal challenges and the importance of the class meetings being "safe and comfortable":

> I felt uncomfortable at various points in the class tonight and I think I finally kind of figured out what it was. There's so much judgment attached to things like whether you shop at Wal-Mart, whether you shop at the co-op and why you make those choices. And I just want to be clear that if people are feeling offended by anything that gets said maybe that's something we should just put on the table to keep it safe and comfortable. As I sit here remembering that I just bought this pen at Target and I looked at the tag and it was made in China... it's really hard. But we've all got our own demons to face in terms of sorting this stuff out. And I just don't want to feel like we feel demonized by each other is all.

Holly also commented in a class meeting mid-semester how reading articles left her in a state of despair:

> We are chasing our tails trying to get everything done that we get done and then reading these articles... and sometimes you just want to say okay, enough is enough, I just suck. I totally suck... and I hate to feel that way.

Later in the semester, Ann commented,

> I'm starting to see how... all the social justice issues are so related and they keep coming up over and over. It's been actually a really good experience for me personally just to talk about all this stuff.

All four teachers in this group also reflected on these ideas at the end of the course. Roxy, in her final course paper, reflected, for example, that she

> became more aware of local and state politics regarding minimum wage.... It bothers me that I have so much when so many have so

little.... The biggest, and perhaps most critical shift in my perception is that cynicism is not a luxury we can afford.

Holly similarly noted in her postinterview that "I'm not the most informed individual and teaching, going through this math class has made me go, well, you know what, maybe I should be a little bit more informed about my community."

Curriculum Development & Revision: Instantiating Goals in Practice

An additional challenge that arose for the teachers, even though they had decided to focus more explicitly on the social justice component of their lessons, was instantiating these ideas into practice. During lesson implementation, teachers had to again focus on *both* the mathematical and the social justice goals of the lesson, and this resulted in the math goals "trumping" the social justice goals. All four teachers were able to implement the lesson in varying classroom contexts (e.g., 90-minute and 45-minute periods, regular and Honors Geometry, Algebra), and they each observed one another teach. The lesson asked students to use local housing data to determine a living wage and to examine the discrepancy between minimum wage and living wage. As the teacher who taught the lesson first explained, "What didn't happen, because I didn't get to that discussion, was we didn't have the impact of, what are we going to do? This isn't right."

None of the teachers, in fact, felt as if in their implementations that they got to enough, or any, of the social justice discussion anticipated for the end of the lesson. Ann reflected on the first two lesson implementations saying, "One thought that I had...is that that's where our comfort zone is. Is in going over this kind of stuff and talking about the mathematics." Dana remarked,

> Going over homework took forever and you would think after observing it twice before, I would make more appropriate decisions.... Maybe that was because I wanted to get to the mathematics there. Maybe that's because that's where we're all comfortable.

Finally, Roxy reflected, "I went down the same rabbit hole we all did with these openings, homework review, because it was just so math dense, just too tempting." Roxy later mentioned that as a group,

> none of us were able to escape the lure of multiple solution strategies to the same problem, unpacking the mathematical proof in

student work, and displaying more than one student solution to a problem [so] we never finished the piece of a living wage is important, not everyone has a living wage, how could we solve that problem mathematically?

It is important to note that though the teachers debriefed after each lesson implementation as a group and had an opportunity to make changes to the lesson plan, they made no significant alterations until all four lesson implementations were completed. Rather, they made small changes to the order of the question asked or chose to make no changes but agreed to be more cognizant of the time issue. Before the final implementation, for example, one teacher remarked, "I don't think there need to be any modifications. The note that I'm giving myself is...how can I do [the warm-up] in a way that gives as many voices to that work and that keeps things moving?" It is also important to note that all four teachers reported they engaged in these discussions (around the social justice goals) with their students the following day in their classrooms and that they completed the lesson as planned.

Discussion: Implications for Research and Practice

A Success Story?

Analysis of the data suggests that engagement in readings and sample lessons around understanding mathematics teaching for social justice and in fundamental components of the lesson study process supported teachers in creating a lesson that incorporated both mathematics and social justice goals. The teachers' lesson, as written, requiring students to use mathematics to examine issues of minimum wage and living wage as seen through housing data in their local community engaged students in both "reading the world" and "writing the world" with mathematics (Gutstein, 2006). The lesson required students to use mathematics to learn about an issue of social injustice, mainly that, as Dana stated, "it's not right that somebody works 40 hours a week and can't afford to live." Additionally, the lesson could engage students in deepening their knowledge of the sociopolitical contexts of their lives, including challenging the notion that this is a tale about individual betterment, or to confront the myth that success is about "pulling oneself up by one's bootstraps." Further, mathematics is used in this way to support students in action toward equity, requesting that students discuss the discrepancy between minimum wage and living wage and brainstorm and share possible conclusions and/or solutions to this issue as part of the lesson plan. Additionally, two of the three homework assignments contained action components, one focused on writing a letter to the editor of a local newspaper and the other on challenging federal regulations.

A Question of Grain Size?

Though the teachers were successful in creating a mathematics lesson for social justice, they acknowledged a tension in negotiating both sets of goals and, as a result, chose to focus more on the social justice component of the lesson during the lesson design stage. This knowledge and concentrated focus was, however, insufficient in supporting teachers in instantiating the social justice goals in their practice. In all four implementations of their lesson, students did not engage with the social justice component as intended, or as one teacher put it, the mathematics "trumped" the social justice.

Close examination of teachers' comments about the mathematical discussions suggest that the teachers were proud that their students engaged in discussions of which measure of center was better and that students asked each other questions such as "Did everybody get this number?" "How did you get that?" Whether or not such comments are representative of "amazing mathematics" is debatable, but it seems that for Holly, Dana, Ann, and Roxy, three of whom were enrolled in a Master's program at the local university that examined research around teaching mathematics for understanding, engagement in the lesson study process also supported their developing practice as mathematics teachers focused on teaching for understanding, and that took precedence for them. Their beliefs about mathematics teaching and learning, namely, about the ideal norms of a classroom focused on learning mathematics with understanding, seemed to mediate the effects of this knowledge in practice.

One thing that comes to mind here is to what extent the teachers viewed this lesson as something to be completed in a single class day. Perhaps teachers felt that since they had planned the lesson initially for one class day that this was the timeframe within which they had to work. Thus, adjusting the lesson to facilitate getting to the social justice component would mean sacrificing some of the mathematical discussion, a negotiation they were unwilling to make. Although the use of lesson study did provide teachers an opportunity to collaborate, to reflect on their lives and their teaching, and to anticipate student responses to the lesson, the focus on a single lesson plan, as implied for these teachers with the term "lesson" in "lesson study," may have limited their conceptions of teaching mathematics for social justice. As such, changing the grain size from the lesson level to a focus on continued integration throughout the school year, or examining one's developing practice across multiple grain sizes (e.g., individual teacher-student interactions, societal, school, classroom, full year, single lesson), may be more appropriate. Professional development could engage teachers with questions such as, What does it mean to teach mathematics for social justice throughout a school year? What does it suggest about my relationships with students?

In postcourse interviews, the teachers' own comments shed light on what such a focus might mean. Ann reflected, "I think there's the thread that connects any social justice issue, lesson. So you know if you had an ongoing series of lessons that you did throughout the year, you might start to see those things coming out." Additionally, Roxy reflected, "Social justice is not just the curriculum or an overarching set of ideals and goals. It is also every tiny interaction with every student, a look, a tone of voice, a choice of word."

More Space for Personal Reflection?

Analysis of the data also suggested that teachers were challenged in wrestling with personal feelings while learning to teach for social justice. Specifically, they worked to reconcile what they were hoping to promote in their classrooms and what they were learning in course readings with choices they made in their personal lives. Researchers who have examined teachers learning to teach for social justice have noted that teachers need to be "willing to examine and deal honestly with [their] own values, assumptions, and emotional reactions to oppression issues" (Bell, Washington, Weinstein & Love, 1997, p. 299). Often this process brings with it feelings of guilt, shame, or embarrassment. Thus, it may be appropriate that in learning to teach mathematics for social justice teachers are provided more of an opportunity to work to identify the biases and assumptions that influence their social identity and to talk through these feelings as they arise, perhaps especially within the specific context of a lesson they are thinking about. One way the lesson study process might support this is through its focus on anticipating students' responses. In thinking about how students might respond to a particular social issue, the teachers could be asked to share and consider their own reactions to these issues. This process may also enable teachers to provide thoughtful response in the classroom when students express similar feelings.

Researcher Reflections

As facilitator of this professional development, I was fraught with questions about my own knowledge. What else must I learn about myself, my beliefs, and my unchallenged assumptions about groups of people? In what ways might I be inhibiting the teachers' abilities to engage fully in this process?

In writing this chapter, additional questions arose. How can I convey the complexity of this situation without sending a message that reifies the status quo? I did not want people to see this study and the teachers' challenges as verification that teaching mathematics for social justice should not be pursued. Should I report about the group's struggles, or write more about what *did* seem to support teachers in learning to teach for social justice? Am I being true to the data if this chapter does not seem to tell the entire story? Thinking about this led to still more questions. Who should teach mathematics for social justice? Anyone? Everyone? What is considered sufficient prior knowledge for effective teaching of mathematics for social justice?

I want to be clear that I know that taking one course, or even an entire preservice teacher education program, is not enough to become an expert mathematics teacher for social justice. And one study is not enough to learn all we can about teachers learning to teach mathematics for social justice. But they might be enough to begin to challenge and further our understanding and our assumptions about the goals of mathematics education and about the knowledge required of teachers negotiating various aspects of this practice.

I leave you with some final thoughts from the teachers in this study:

Roxy

I learned that teaching mathematics is in my blood, and teaching mathematics for social justice is not far off...it's going to affect how I address groups of students, whole classes of students and the biggest thing is that I am not going to pretend math is not political anymore.

Holly

I learned what it means to teach mathematics for social justice and also, to some extent, that I have been engaging in this practice already... [my colleagues and I] are already discussing how to use this lesson to start next year.

Ann

I see it as possible for me [to teach math for social justice].... It might make me more, you know, have my ears more open to things kids are talking about in terms of wanting to connect any other lesson to things they're already thinking about...and I think, you know, I can see myself talking to all three of these women next year and...I think we can do more with this.

Dana

It definitely made me want to pursue social justice teaching, and has made me think more about my students' experiences in bringing that into the classroom to enrich their mathematical learning.

Notes

1. All participants' names are pseudonyms.
2. Throughout this chapter, I am intentionally not listing certain teacher characteristics, such as years of teaching experience and school affiliation, to protect anonymity.

CHAPTER 12

Keeping the Mathematics on the Table in Urban Mathematics Professional Development: A Model that Integrates Dispositions toward Students

JOI SPENCER, JAIME PARK,
AND ROSSELLA SANTAGATA

Which types of teacher knowledge are most important in shaping instructional practice and student learning opportunities in the urban mathematics classroom? Researchers in the field of mathematics education assure us of the role of mathematics content knowledge and pedagogical content knowledge (Hill, Rowan & Ball, 2005; Ball, Thames & Phelps, 2008; Ma, 1999). This work has pushed the field to reconceptualize the types of knowledge needed to provide effective mathematics instruction.

Yet, as compelling as this body of work is, efforts in urban mathematics classrooms tells us that the story of which type of knowledge is needed to teach mathematics is more complex still. On the basis of the results of a three-year study of a video-based mathematics professional development (PD) in one of the largest urban school districts in the country, this chapter examines and explores the interplay between mathematics content knowledge, pedagogical content knowledge, and teachers' disposition toward their mathematics students.

A reconceptualization of disposition as teacher beliefs about and stances toward their students as doers of mathematics is offered, as well as a discussion on the constraints, affordances, and difficulties of theorizing around and preparing PD that addresses this construct. Analysis of video data is used to augment and support this discussion as we explore how to better prepare teachers for effective instructional practice in urban mathematics classrooms.

We begin this chapter with a discussion of Pre-Algebra Learning (PAL), a two-year mathematics PD program implemented in a high-poverty, racial and linguistic minority school district. The PD program was based on results from the Third International Mathematics and Science (TIMSS) Video Study. The TIMSS study revealed that mathematics lessons presented in high-achieving nations were mathematically sound and cognitively rich, requiring students to engage deeply with mathematical concepts and skills (Hiebert et al., 2003). In U.S. lessons, on the other hand, solution methods were reduced mainly to procedures to be followed with little or no connection to underlying concepts. PAL was developed with two hypotheses in mind that might explain TIMSS findings:

1. U.S. teachers do not possess a deep understanding of the mathematics they are asked to teach.
2. Teaching mathematics for understanding is not consistent with the tradition of school mathematics in the United States Thus, U.S. teachers seldom have the opportunity to observe examples of teaching in which the cognitive demand of problems is maintained (Van Es, & Sherin, 2008; Hufferd-Ackles, Fuson & Sherin, 2004).

PAL included opportunities for teachers to deepen their own understanding of key concepts of the curriculum they teach, improve their knowledge of ways students understand the content, and learn about instructional strategies that can be used to maintain the cognitive demand of the problems they pose. The PAL study specifically tested the effects of the PD program on teachers working in urban middle schools in the United States.

Professional Development Structure and Content

The PAL program consisted of three modules, each targeting a key content area of the California sixth grade curriculum (fractions, ratio and proportion, and expressions and equations), and its core concepts (Santagata, Kersting, Givvin, & Stigler, 2009). Teachers met face-to-face in groups of eight to ten led by one or both of two facilitators, each having a strong background in mathematics and several years of teaching experience. Each teacher was provided a laptop with Internet connectivity. Video-based analyses were supported by Visibility, a multimedia platform developed by LessonLab.[1] Each module was structured into three main folders: (a) Content Exploration, (b) Lesson Analysis, and (c) Link to Practice. Within each folder, numbered pages guided teachers through a series of video-based analysis tasks, each containing a few questions for them to answer. Teachers were asked to watch preselected video segments and type their responses to the questions into a textbox. Some questions also required teachers to point the reader's attention to specific moments of the video. A feature of the software allowed them to click on a button that inserted in their text a time stamp corresponding to a moment of the video they had chosen to cite. Written responses were saved on a server accessible by both the facilitator (during the PD sessions) and the researchers (who later were able to analyze them).

Independent work on the laptops was interspersed with whole group discussions led by the facilitator, who sometimes projected participants' written responses onto a screen. Teachers spent roughly half of the time working independently on their laptops and half of the time working and discussing as a whole group. After an initial introductory meeting, teachers met six times in the course of the school year, spending two full days working on each module. Content Exploration and Lesson Analysis were each addressed for a full day, usually a week apart. A teaching window followed, after which teachers met at their school sites for one hour to share their teaching experiences as part of the Linking to Practice phase. The modules were distributed across the school year so teachers would participate in the PD sessions on a particular topic area immediately before they were to teach it to their students. What follows is a description of the content and structure of the modules.

Content Exploration

Content Exploration was aimed at deepening teachers' understanding of core mathematics concepts and was accomplished through a combination of written documents and video. Instead of being asked to watch a videotaped classroom lesson, here, teachers were exposed through video to a mathematics-focused discussion among other teachers, led by a mathematics educator. This had the dual purpose of (a) providing a dynamic setting for teachers to learn mathematics concepts that would make the task more engaging than simply reading mathematics content documents, and (b) creating an atmosphere in which teachers would feel comfortable sharing doubts they themselves may have on mathematics concepts.

PAL teachers watched selected segments of the videotaped discussion and posted online answers to questions aimed at fostering their conceptual understanding. Individual teachers' responses were then shared in a group discussion. Occasionally, concrete materials were provided to the teachers to enable them to engage in the same activities as the videotaped teachers.

Lesson Analysis

For each module, the second day of PD was dedicated to Lesson Analysis. Teachers began by solving a rich problem that made use of one or more of the core concepts studied the day before. They then studied a lesson plan that incorporated the rich problem and watched the video of the lesson in which the rich problem was taught. This videotaped lesson provided teachers with a model for engaging students in conceptual thinking. Lessons were filmed in the teachers' district; thus, the students portrayed in the video were from the same population as the participating teachers' students. During this phase, teachers answered a series of questions aimed at the analysis of students' learning and understanding as evidenced in the video and in samples of students' work. At the end of the day, teachers discussed ways the lesson plan could be improved and proposed modifications before going back to their classrooms to teach the resulting lesson to their students.

Link to Practice

During this phase, teachers taught the lesson they had analyzed and participated in a facilitator-led one-hour meeting at their school sites. This phase was aimed at facilitating the application of what was learned during the PD sessions to the teachers' daily practices. At the meeting, teachers were asked to share with their colleagues samples of student work from the lesson they had taught. Each person took a turn and summarized his or her experience teaching the lesson and shared student work to discuss both aspects of the lesson that went as planned and aspects that did not.

Study Participants and Methods

Sixth grade teachers from five low-performing Title I (i.e., having a poverty rate of 50 percent or higher) inner-city middle schools participated in the PD program. Two of the schools followed a regular September to June calendar. The remaining schools followed a year-round calendar, with three or four groups of students rotating throughout the calendar year. All five schools were very large in size, accommodating on an average 2,270 students in Grades 6–8. Each school included approximately 20 sixth grade classes, and most sixth grade teachers taught mathematics and another subject matter (usually science) to three groups of students. The student population was predominantly Hispanic (from 62 to 95 percent) and Black (from 18 to 38 percent). Between 30 and 47 percent of the students at each school were English language learners. Student mathematics achievement in these schools, as measured by standardized tests, was among the lowest in the state, with only 6.2 percent of students on an average reaching a proficient level.

The PAL program was implemented for two consecutive years and was made mandatory by the district. During the first year, a randomly selected half of all sixth grade teachers at these five schools attended the PD. The remaining teachers were included in a "no-treatment" control group. During the second year, all teachers participated in the program.

Program effectiveness was investigated through an experimental study. Measures included (a) a teacher survey to study teacher knowledge of mathematics for teaching prior to the PD and at the end of each year of implementation,[2] (b) the videotape of one lesson per teacher each year, in which teachers were asked to teach a given problem, and (c) student district-mandated quarterly assessments and the state standardized test administered at the end of each year. Multiple contextual factors hindered the success of the program and made the completion of the research study challenging. Yet, findings from the first year of implementation indicated a significant impact of the program on performance on the quarterly assessment of those students whose treatment teachers had reached a certain level of content knowledge (compared with the performance of students whose teachers in the control group reached a similar level of content knowledge). Findings from the second year of implementation also indicated positive improvements in both teacher knowledge and student learning. Issues of missing data did not allow a review of the findings in terms of their statistical significance. For more details on the study design and findings, see Santagata and her colleagues (2009).

Telling the Full Story

The results of PAL revealed that participating teachers who already had a certain level of content knowledge were able to benefit most from the PD. As other studies have shown (Carpenter, Fennema, Peterson, Ci-Pang, &Loef, 1989; Hill, Rowan, & Ball, 2005), teachers with higher levels of pedagogical content knowledge are best positioned to leverage this knowledge to the benefit of their students. Yet, for those of us engaged in the PAL study, it became clear that this finding was merely the backdrop to a much more complex set of stories that shaped mathematics learning and instructional practice at the participating schools. These stories have pushed us to consider which measures (beyond content knowledge and pedagogical content knowledge) we ought to consider when working to make sense of the mathematics teaching and learning in contexts serving high populations of poor and minority students. In our work in this context, we witnessed a number of unique challenges. These included (a) extreme accountability pressures from the district office and state, (b) an excessive number of assessments given to students due to the schools' past poor performance on standardized exams, and (c) numerous, disparate PD workshops that teachers were required to attend. These challenges greatly constrained the PAL PD program. But beyond these technical constraints we encountered a set of troubling norms toward students that encouraged us to rethink the work of PAL more specifically and mathematics PD more generally. While we continue to debate over an accurate term for these normative constraints, we feel that it is essential to bring them to light as a part of the findings embedded within the overall PAL study.

Troubling "Dispositions" toward Students

Evidenced through the numerous discussions held during our PD workshops, teachers' classroom practices, and teachers' active resistance to our suggestions to deepen the mathematical content in their classrooms, many PAL teachers demonstrated a troubling set of beliefs about and stances toward students as learners and doers of mathematics. As our study progressed, we observed that what made a teacher engage or not engage in our PD had to do with the teacher's beliefs about his or her students' abilities to do the rigorous work that we prescribed. Furthermore, as we witnessed PAL teachers' mathematics instruction, we noticed another connection. Engaging students in the rich PAL lessons was not only connected to content knowledge, but to what a teacher decided to do with his or her content knowledge. Most important to us, as mathematics educators, was the manner in which these norms shaped the actual classroom practices (the mathematics tasks given, the sequencing of mathematics tasks, the expectations placed on students for successfully completing those tasks, the nature and norms for mathematics work, the norms for what is accepted as mathematical work and mathematical accuracy) of PAL teachers. For example, in reviewing PAL videotaped lessons, we found that teachers spent large percentages of class time on "warm-ups." We found that many of the warm-up problems presented to students were mere replications of the rich PAL problems that students would subsequently work on. Presenting these "warm-ups" before the actual PAL problems often resulted in teachers giving away the solution methods or procedures for the richer problem that was to follow. While not taken up in this chapter, these classroom practices are explored in Spencer (2009). Furthermore, of great importance to us was the manner in which these dispositions manifested themselves in the classroom talk of PAL teachers. As classroom teacher talk provides a view into classroom practices, we devote the next section of this chapter to presenting and analyzing the talk of selected PAL teachers as captured by videotaped classroom episodes. We share these observations in hopes of engaging our readers in a conversation about supporting mathematics learning and teaching in schools that have large populations of racial and linguistic minority students.

Examining Teacher Dispositions Through Four Cases of Classroom Teacher Talk

Classroom teacher talk is one way to examine teachers' disposition toward students. We share four cases of classroom teacher talk in order to understand the different ways teachers might talk to students and the implications of this talk for student opportunity to engage in rigorous mathematics.

In the second year of the PD, all teachers were asked to teach the Carnival Problem:

> A carnival game has ten plastic fish in a pond. Without looking, you use a fishing pole to catch a fish. You do this two times. In order to win, you must get a red fish each time. The pond has two red fish, four yellow fish, three blue fish, and one green fish.

Analyses of videotaped lessons of the Carnival Problem provide the basis for our examples. We reviewed all tapes and identified four that we think best represent the range of classroom teacher talk. The first three examples illustrate teacher talk that gets in the way of engaging students in rigorous mathematics. The teachers in these three examples underestimate students' potential to solve rich mathematics problems. However, the fourth teacher illustrates teacher talk that is about doing mathematics.

> Mr. A: *What is fishing?* Mr. A spent over 20 minutes in a 60-minute class period discussing fishing. Selected dialogue from the segment is included below.
> Mr. A: Okay, hold on a second. I'll get you. Fishing. How many of you have been fishing? Raise your hand.

★★★

> Mr. A: You might see a fishing competition on ESPN, or maybe on the Outdoor Life Network. What is it that all sports have so that we know how to play?
> Student (SN): Rules.

★★★

> Mr. A: So you might ask for?
> SN: Catfish.
> Mr. A: Catfish, or—
> SN: Salmon.
> Mr. A: Or salmon. Or what else?
> SN: Shark.
> SN: Dolphin.

Mr. A: Oh, a dolphin's not a fish and we wouldn't eat that anyway.
SN: Oysters.
Mr. A: Oyster's not really a fish.
SN: Sushi.
Mr. A: Sharks, tuna, cod. Not all fish are the same.

★★★

The teacher asks students to draw pictures of fishes.

Similar to this example, many of the teachers in our PD discussed the context of the problem prior to letting students work on it. Lubienski (2008) suggests that teachers need to be mindful about using contextualized mathematics problems with low-income students because in a study she conducted, she found that they became deeply engaged in the context and missed the intended mathematical ideas in the problem. However, teachers' false assumptions about their students' knowledge may lead to classroom teacher talk that is unrelated to the mathematics at hand. Too much nonmathematical talk may deteriorate the intention of the mathematical task and keep students from engaging with the mathematics, as seen in the example presented earlier.

> *Ms. R*: "You might think of something. That should not be hard." In the second example, the teacher solves a similar problem to the Carnival Problem first. She multiplies two fractions to get the solution. Next she poses the Carnival Problem to her students and asks for a method different than multiplying two fractions.
>
> *Ms. R*: What are some of the other ways you might represent your probability? Who can think of another way you might represent your probability?
>
> *Ms. R*: How?
>
> *SN*: Line graph?
>
> *Ms. R*: A line graph? A line graph shows change over time, and you might have some changes there, but, uh, what else?
>
> *Ms. R*: I don't think a line graph or a bar graph would be perhaps your best choice there.
>
> *Ms. R*: You guys are really creative when it comes to thinking of ways to outwit the teacher. Think of ways to do it better than I did.
>
> *Ms. R*: That should not be hard. The only hard part is I get to judge whether it's better than what I did.

Ms. R: Now, instead of laughing at what everybody else does Fernando[3], why don't you do something creative? You're one of my best math students.

Ms. R: Come on. Edgar. Come on. Think of a way you could represent this, pictorially, and you don't have to, uh, you don't have to use a graph. You don't have to use a table, but think of a way you could represent your probability in a picture maybe.

Ms. R: Fish are easy to draw. I drew these, and it's been years since I took art. All you have to do is make a sideways eight and then draw the tail.

Alternative methods to finding probabilities of compound events are specialized and difficult for anyone to figure out on their own without sufficient understanding of the mathematics. In the excerpt, there is little opportunity for students to see that particular mathematical methods are connected to particular mathematical ideas. There are no bases for their guesses, and the teacher does not guide them to an appropriate solution strategy. Ms. R tells students to be "creative" and minimizes the complexity of the problem by telling them to draw fishes.

This case illustrates poor instructional decisions that limit opportunities for students to engage with the mathematics. However, without careful analyses of teacher practice, the lack of student participation may make one think of students as unable to handle the rigor of the mathematics.

Ms. T: *"Right. Very good, class! Excellent! You are so smart."* Here, the teacher, Ms. T, focuses her students on the basic *operations* of mathematics as opposed to the *understanding* of the mathematical context.

Ms. T: Well, representation is just another way of showing something, but what word up here does Ms. T keep telling you over and over?

SN: Operation.

Ms. T: Operation. And what are the operations?

SN: Multiply, divide—division, addition, and subtraction.

Ms. T: Right, and Ms. T says that over and over all through sixth grade. There's only four. What are they?

SN: Add, subtract, multiply, and divide.

Ms. T: Right. Very good, class.

Integrating Dispositions toward Students

Ms. T continues to focus on the procedural aspects of the problem throughout the class period. The students have been working on multiplying $2/10$ and $2/10$, and $2/10$ and $2/9$. Ms. T makes each group go up to the board and present the same work. Multiplying fractions and converting them to decimals and percents are covered under Grades 2–4 state standards.

> *Ms. T:* I want Marisol and Martiza and Victoria's poster next. Bring it up here.
>
> *Ms. T:* That's great work, Omar. Are you guys brave enough to come up here for me? Oh, you're going to make me be on the spot?
>
> *Ms. T:* Maritza's brave. Okay, look at this. Check this out. These kids worked on that challenge problem, and I want you all to show them some respect right now by learning from them. Victoria, you're brave enough, come on up.
>
> *Ms. T:* Okay, come on, Marisol. All right. Okay, Maritza, the only brave one. Let's go over it together.
>
> *Ms. T:* Oh, come on, bring it up. Let's share it. I'm so excited. I'm so excited somebody got it. Let's pass these back.
>
> *Ms. T:* Thank you, Maritza. Good job. You're so smart.

This case illustrates ways in which teachers can be so concerned about their students' self-confidence that students are precluded from participating in meaningful mathematical thinking and discourse. This concern is grounded in a belief that to help poor children, we simply need to be nice to them. However, the nicest thing that a teacher can do is to prepare her students to meet the academic challenges they will face. Instead of increasing student confidence and ability, Ms. T demonstrates low expectations and false generosity.

> *Ms. O: Doing mathematics.* Ms. O provides an example of a teacher who engages her students in doing mathematics by providing the necessary guidance and expectations for students. As soon as the bell rings, the mathematical goals are made explicit.
>
> *Ms. O:* We're going to do different—differentiate between compound events with replacement and without replacement. It's in your class work as usual, so we can—those are the objectives, our goals today. We're going to represent the outcomes of compound events in various ways. We will interpret the representation in order to express the theoretical probabilities that

we have done in—before. And make progress toward understanding the relationship between representation and the actual event.

Then she has a student read out a word problem similar to the Carnival Problem. Ms. O tries to address any language needs students may have.

>Ms. O: Who said that. Very good. That's our new student, Kinea? Randomly. They're chosen—that's an important word. Randomly.
>Ms. O: They're chosen randomly. What does choosing randomly mean? Yes, Alex?
>SN: Without—oh, I thought you said like randomly.
>Ms. O: Randomly, random.
>SN: Yeah. Uh, without order.
>Ms. O: Without order.
>SN: Um, the chance.
>Ms. O: Chance. Probability is chance. What's the chance. Can you give an example? Can you give an example? Yes, Alex?

Next, Ms. O brainstorms with her students the different solution pathways for figuring out this problem.

>Ms. O: Okay. So, we could use tables. Using tables, which is what our standards said. We need to know, how to use tables, grids. What's another way I could do it?
>SN: Diagrams?
>Ms. O: Hmm?
>SN: Diagrams?
>Ms. O: Tree diagrams. Very good. Tree diagrams.
>Ms. O: Tree—did any group use tree diagrams?
>SN: Oh, we did.

As students work on the Carnival Problem, Ms. O walks around and supports students by pointing them to key information, evaluating solution methods, and asking for student consensus.

>Ms. O: What did you say? Okay, you need to think—think about the first one, then the second one. The first time, you pick—
>SN: So, then you pick it.

SN2: Okay, and there's ten and—
SN: It's one out of nine.
Ms. O: Claudio? Let's listen to Claudio. Yes? What did you say?
SN: Um, so before you pick it, um, it's two out of ten, and when you pick one and you get it, then the probability is, um, one out of ten
Ms. O: Still one out of ten? Claudio, what do you say?
SN: One out of nine.
Ms. O: One out of nine. Okay, you guys. Why do you say one out of nine?
SN: Because you take one fish away.

This case illustrates a teacher who believes that students can engage in meaningful mathematical thinking and discourse given the right kind of support. Students were repeatedly asked to make sense of the mathematics. And they were able to because of the deliberate instructional support Ms. O decided to provide for her students that gave them access to the mathematics.

Four Cases of Teacher Talk and Students' Opportunities with the Mathematics

We have examined four cases of teacher talk. To further examine the kinds of opportunities students in each of these teachers' classrooms had to engage in the mathematics, each lesson was coded using the following categories: (a) Number and Type of Solution Methods, (b) Probability Reasoning, and (c) Logical Reasoning.

Number and type of solution methods. A variety of solution methods might have been discussed in the lesson. We were interested in these discussions without differentiating whether the discussion was initiated by the teacher or a student. The solution methods include different ways to find probability of compound events. Drawing pictures can help one to figure out the probability of a single event, but this method is not viable for compound events and therefore not included in our codes. Table 12.1 shows that Ms. O's class is the only one that discussed solution methods other than multiplying fractions.

Probability reasoning. For the Carnival Problem, probability reasoning entails understanding (a) Outcomes; (b) Probabilities as Ratios, Proportions, Decimals, and Fractions; and (c) Independent and Dependent Events. For each idea, criteria were developed to rate each

Table 12.1 Number and type of solution methods discussed by teacher and/or students

Teacher	Number of Solution Methods	Types of Solutions
Mr. A	1	Multiplying Fractions
Ms. R	1	Multiplying Fractions
Ms. T	1	Multiplying Fractions
Ms. O	4	Multiplying Fractions
		Generating Empirical Probabilities
		Outcome Grid
		Outcome Tree Diagram

Table 12.2 Probability reasoning discussed by teacher and/or students

Teacher	Probability Reasoning		
	Outcomes	Probabilities as Ratios, Proportions, Decimals, and Fractions	Independent and Dependent Events
Mr. A	criterion not met	criterion not met	criterion not met
Ms. R	criterion not met	criterion partially met	criterion partially met
Ms. T	criterion not met	criterion partially met	criterion not met
Ms. O	criterion fully met	criterion fully met	criterion fully met

lesson. Table 12.2 shows whether each criterion was partially or fully met or not met at all during the lesson.

Logical reasoning. We describe logical reasoning for the Carnival Problem as (a) Understanding Reasonableness of Probabilities, (b) Justifying Choices with Logical Necessity Statements, and (c) Pursuing and Describing Problem-Solving Strategies. Again, criteria were developed for each idea and were used to rate each lesson. Table 12.3 presents these results.

Except for Ms. O, all teachers used substantial amounts of classroom time on nonmathematical talk and/or mathematical talk that reduced the cognitive demands of the Carnival Problem. This is further reflected in opportunities students had in each classroom to engage with solution methods, probability, and logical reasoning associated with the Carnival Problem.

Concluding Thoughts on Teacher Classroom Talk

Teachers talk to students in particular ways on the basis of their knowledge and assumptions about the mathematics, the teaching of

Table 12.3 Logical reasoning discussed by teacher and/or students

Teacher	Logical Reasoning		
	Understanding Reasonableness of Possibilities	Justifying Choices with Logical Necessity Statements	Pursuing and Describing Problem-Solving Strategies
Mr. A	criterion not met	criterion not met	criterion not met
Ms. R	criterion partially met	criterion partially met	criterion not met
Ms. T	criterion not met	criterion partially met	criterion partially met
Ms. O	criterion not met	criterion fully met	criterion fully met

mathematics, and the students. The ways in which teachers' content and pedagogical content knowledge impact student learning opportunities have been highlighted by many (Ball, Thames, & Phelps, 2008; Hill et al., 2005). Here, as mentioned earlier, we focus on a different aspect of teacher knowledge, what we call dispositions toward students. Teachers bring to the classrooms assumptions about their students that can have profound effects on opportunities students are given to engage in mathematics learning in rigorous ways. The biases against students are subtle, as we have illustrated through the cases.

Defining Disposition

Our numerous observations, of both instructional practices and teacher talk, pushed us to consider that something beyond teachers' initial mathematics content and pedagogical content knowledge was shaping our PD work. We employed the term "disposition" in an effort to define this "something" that we witnessed during the PAL study. Examining the literature on disposition in the context of mathematics teaching and learning, we found very little, if any, that spoke to what we were attempting to define as a particular set of teacher beliefs about and stances toward their students as doers of mathematics. Instead, much of it dealt with disposition toward the *teaching* and *learning* of mathematics.

Seeking to understand the affective factors associated with some of the individual differences in mathematics achievement, Fennema and Sherman (1976) developed nine Mathematics Attitude Scales. These scales measured affective factors, such as confidence, anxiety, and motivation, hypothesized to be associated with math achievement. The National Council of Teachers of Mathematics (NCTM) takes up the issue of disposition in their Standards documents (NCTM, 1989,

2000). "Students' mathematical dispositions," they state, "are manifested in the way they approach tasks whether with confidence, willingness to explore alternatives, perseverance, and interest and in their tendency to reflect on their own thinking" (NCTM 1989, p. 233). Invoking a similar set of ideas, NCTM (2000) describes "math power" as a combination of ability and attitude—a disposition to question, explore, and engage in significant math problems. The authors of *Adding it Up* (National Research Council, 2001) devote discussion to the issue of "productive disposition." This productive disposition, the authors posit, includes students' sense of themselves as mathematically capable as well as intellectually autonomous in the field of math. Similarly, the work of Yackel and Cobb (1996), with reference to the *Professional Standards for Teaching Mathematics* (NCTM, 1991), takes up the notion of development of a mathematical disposition. In their work, the authors investigate the issue of mathematical disposition as, "how students develop specific mathematical beliefs and values, and consequently, how they become intellectually autonomous in mathematics" (p. 458). DeCorte (1995) pushes his readers to explore how the mathematics education research field might move toward a dispositional view of mathematics learning. Reviewing works on mathematics affect (see for example McLeod, 1994), DeCorte argues that mathematics learning is not only a cognitive act, but an act of a learner's beliefs, attitudes, and emotions.

In many of the previous studies, analysis is centered on the student. There are also studies that focus on teachers as agents in shaping the math disposition of their students. For example, DeCorte (1995) asks, "How might learning environments be built to support the mathematics disposition of students?" (p. 40). Hart (2002) developed the Mathematics Beliefs Index (MBI) in an effort to evaluate teacher education programs' abilities to promote teacher beliefs and attitudes that support the philosophies underlying current math reforms.

The impact of disposition on mathematics instructional practice is less prominent in the literature. Ball (1996) establishes this link, stating,

> What teachers bring with them (into the classroom) is not purely cognitive, for they also bring commitments about how to act with different students, a sense of themselves as helpful and effective, and feelings about certain kinds of classroom environments. These, too, influence their interpretation of and disposition toward the mathematics reforms. (p. 500)

Similarly, Cabello and Burstein (1995) write, "Instructional practices embody teachers beliefs, which are a reflection of their own experiences and background" (p. 143). Research is needed to uncover the ways in which the experiences and backgrounds of teachers impact their mathematics instruction, and their students' learning. There is a critical need for such research centered on the urban, ghetto, and central city school context. Given the mathematics education community's stated concern with the racial achievement gap in mathematics, it is imperative that researchers begin to grapple with some of the more salient, often invisible, reasons that this gap exists and persists.

Beyond documenting deficit dispositions, and their impact on mathematics instructional practice, there is a need for PD that effectively addresses such views and supports teachers in developing successful mathematics practices in their urban classrooms. PD should engage teachers in "critical" versus "technical" reflection (Zeichner & Gore, 1990). We find promise in several lines of work, including Gau (2005); Foote (2006); Dunn (2005); Battey (2004); Park, (2008); Koehn, (2009); and Ho (2009). Each of these young scholars has worked to engage urban mathematics classroom teachers in critical versus technical reflection, pushing them to consider their preconceived notions about and dispositions toward their students.

Discussion: Keeping the Mathematics Central

How is it that one of our teachers was able to provide high-level mathematics instruction and learning opportunities to her students while so many others were not? What can we learn from her example as we work to improve teaching and learning in our specific context? According to the TIMMS study (Stigler & Hiebert, 1999), students in high-scoring countries are given many classroom opportunities to connect mathematical ideas in complex mathematics problems. Those teachers, like Ms. O, who are able to present complex mathematical problems to their students and require them to make connections between mathematical ideas, will be more successful than those who are not able to do this. In the context of inner-city schools, serving large percentages of poor students, racial minority students, and linguistic minority students, PD programs must support teachers in "keeping the mathematics on the table" (Elliot, Lessieg, Kazemi, & Kelly-Peterson, 2009). In other words, teachers must be able to push past deficit thinking and detrimental dispositions toward their students and be firmly committed to the notion that every child has the intellectual ability to think deeply and critically about mathematics. The cases analyzed here have highlighted for us something that PAL failed to do: Engage teachers in discussions about their dispositions toward students from various backgrounds. Teachers in urban schools, particularly those serving high populations of poor, racial and linguistic minority students, are particularly susceptible to deficit dispositions that do not support instructional improvement. More specifically, because deficit views about poor and minority communities are embedded in the American Discourse (of which teachers are a part), these ideas are brought into the classrooms that they teach in. These beliefs are a part of the larger American mythology/belief in meritocracy, namely, if you work hard, then you will "succeed." Conversely, if you (or the community that you live in) has not succeeded then it is because you have not worked hard. Beliefs that students and their communities are lazy, lack motivation, and do not care about school predominate. Because these dispositions are pervasive, dismantling them requires deliberate, conscious effort. Our lack of attention to these realities, we argue, served to stymie our work in real ways. If we were to design again a PD program similar to PAL, we would make central the belief that learning is a function of instructional practice, that instruction can and should be constantly improved, and that students in urban settings are also capable of doing complex, mathematical work.

Conclusion

There is something unique about doing work in schools serving large populations of poor, racial and linguistic minority students. Anyon (1997) describes this phenomenon as "ghetto schooling." Kozol (2006) provides vivid pictures and compelling stories of what it means for students to learn in places that are physically dilapidated, overpopulated, and underresourced. Those who have done work in these spaces can attest to the truths in Anyon's and Kozol's work. They also know that accompanying the physical attributes of these schools is an even more overwhelming and degenerative *ethos*. It is not an ethos of exasperation because exasperation connotes that one has exerted effort to change things. Rather it is an ethos of acceptance and tolerance of mediocrity. Low expectations are so embedded into these spaces that they are no longer noticed. No one questions the gross inequities present because no one sees them as out of the ordinary. The practices and processes are normalized. They are part of the American culture of educating poor children, Brown children, and Black children. This is not necessarily an indictment of the teachers and staff who work in these spaces, as many of these professionals are under immense and unrealistic strains and pressures by their school boards and superintendents' offices to produce the same achievement results as those who work in non-ghetto schools.

While these negative dispositions and ethos shape the work of math education researchers, specifically the work done in schools serving low-income racial and linguistic minority students, they are rarely taken up in the work and research of mathematics education. The lack of theorizing about mathematics teaching and learning in these spaces strikes us as a gaping omission in mathematics education research. In the midst of conducting the PAL study, we found little research to speak to what we were witnessing. More specifically, we found a lack of research discussing how low expectations and deficit views of students impact the actual mathematics instructional practices inside classrooms. Questions that we continue to struggle with include the following:

1. What is the nature of mathematics instructional practice in schools serving low-income racial and linguistic minorities? Which instructional practices are resisted (and why) in such schools?
2. What is the nature of mathematics work and tasks in classrooms serving low-income racial and linguistic minority students? What rationales are given for the nature of this work?
3. Which mathematics topics are taken up in classrooms serving low-income racial and linguistic minority students? What rationales are given for the topics that are and are not taken up in such classrooms?
4. What successful models exist of mathematics instructional practice, teaching, and learning amongst schools serving low-income racial and linguistic minority students?
5. What successful models of mathematics PD and teacher education exist that address the context of schools serving low-income racial and linguistic minority students?

Beyond these questions, we would like to see greater acknowledgment of the unique nature of improving mathematics teaching and learning opportunities in schools serving low-income racial and linguistic minority students. Honoring this work, we argue, begins with abandoning the notion that the work of mathematics education research is neutral and context free. We learned, during the work of PAL, that it is not. Could we have had a great impact on teaching and learning in the PAL classrooms if we had devoted more time to understanding and dismantling the dispositions that shaped the instructional practices and resistance that we saw? Could we have had a greater impact if our work was modeled on projects that had been shown to be successful in the specific context of schools serving low-income racial and linguistic minority students? Quite possibly the answer to these questions is yes.

In a recent symposium of leading mathematics education PD researchers, one team of presenters, responding to the research presented in this chapter, stated, "We [as a research community] need to find ways to keep the mathematics on the table" in urban math PD (Elliot et al., 2009). They went on to state that in our PD efforts, the mathematics, "which so easily falls off the table," must be central. In this chapter, we have shared our efforts, struggles, and questions as urban mathematics education researchers and professional developers seeking to provide rich opportunities for teachers and students. In our work we found that "keeping the mathematics on the table" was a complex endeavor. Professional developers and researchers must honor this complexity by thinking hard about the unique nature of the urban minoritized context. Business as usual will not suffice. Attending to those issues and dispositions that keep rich mathematics out of the hands of the most needy and vulnerable students must become more central in our work. Like our fellow mathematics education researchers, our goal is to drastically improve the mathematics learning opportunities of poor, racial and linguistic minority children through every means possible.

Notes

1. LessonLab is a research center developed by Jim Stigler devoted to the use of video as a tool for improving teaching. The researchers at LessonLab conducted the TIMSS Video Study.
2. The survey consisted of 38 multiple-choice item stems, which resulted in a total of 52 item scores, organized into three sub-scales, each targeting one of the three focus content areas. The majority of items were drawn from an item bank developed by Deborah Ball, Heather Hill, and colleagues at the University of Michigan as part of the Learning Mathematics for Teaching project (LMT).
3. Pseudonyms are used for all students to protect their identities.

Commentary: Part II

Megan Franke and Kyndall Brown

As we look across the chapters in the second section of the book, a number of themes emerge, themes much related to those that arose in the first section of the book. We find that the authors continue to work to elucidate the details of professional development practice that address equity, support teachers to figure out how to make sense of equity within classroom practice, and understand what influences teacher learning and practice. Three themes emerged for us as we read and discussed these chapters: leveraging existing approaches to address equity, supporting teachers to see what students *can* do, and negotiating the design and evolution of professional development.

Leveraging Existing Approaches

These chapters brought to the foreground the power of leveraging existing approaches and structures to support equity. A number of the professional development endeavors drew on existing classroom practices or approaches to professional development and embedded within them a particular focus on issues of equity. Hand and her colleagues built upon the work of Complex Instruction (CI), using it as a way to help teachers think about who was participating, why, and how the teacher played a role in orienting students to participate. Battey and Chan used Cognitively Guided Instruction (CGI) professional development already under way as a context within which to focus on equity, particularly supporting African American students.. Bartell employed lesson study to engage teachers in attending to both mathematical and social justice goals and in the process provided a space for teachers to interrogate their own positionality. Pitts Bannister and colleagues leveraged a think aloud protocol as support for not only developing further content knowledge on the part of the teacher, but for also supporting that teacher in working with students, opening doors to knowing students in new ways.

As we reflect on the benefits of such an approach to professional development, we see that embedding the focus on equity within existing practices allows teachers to benefit from the existing approach itself and supports teachers to use what they have learned to make sense of equity. The existing approaches, of course, need to provide the structure and support for issues of equity. In the case of Hand and her colleagues, CI in conjunction with a focus on equity provided for teachers a pedagogical tool that they could use to think about how to structure opportunities for student to participate. It allowed the teachers to consider the principled ideas at the base of CI in a way that was connected to issues of equity. Within CGI, teachers learned about the development of students' mathematical thinking and used those ideas in their practice, becoming teachers who saw themselves as teachers who listened to students and built upon their mathematical thinking, focusing particularly on what students *could* do. These same teachers then were able to take these understandings and consider what this meant for low-income students of color, and do so in a way that took advantage of the relationships they had built with the professional developers.

The existing approaches provided ways for teachers to make sense of equity within the context of ongoing practice. One potential danger to this approach is that equity could become lost among the ongoing professional development work. This was not the case in the examples in this book. Here, equity was central; the professional developers leveraged existing approaches, helping teachers make sense of equity, use it within their practice, and do so in a way that made sense to them.

Focusing on What students Can Do

Common across the approaches to professional development was the focus on supporting teachers to see what students *can* do. As noted in a number of the chapters, teachers' expectations and assumptions can be a barrier from engaging in practices that support each student learning. We spend a tremendous amount of time in schools identifying and discussing what students don't know and cannot do: middle school students can't do fractions, high school students don't know their basic skills, and so on. Challenging these assumptions, often based on previous interactions with students, is difficult. These approaches to professional development demonstrate that focusing on what students can do orients teachers toward a different kind of classroom practice: a different way of asking questions, of noticing what students are doing and why, and of wondering what they may have missed. Focusing on what students can do reminds teachers that there is a place to build from. While students may not understand all aspects of the mathematics we are looking for, the role of the teacher is to leverage what students do know and build upon that. We recognize that focusing on what students do know becomes more difficult as students get older; however, these chapters reveal that it can have tremendous payoff. Ultimately, this idea leads teachers to treat students as sense makers with potential. In cases like CGI and the think aloud protocol, it also provides teachers with a language and structure for talking about students' mathematical thinking so teachers can continue to build their knowledge.

Negotiating Who Drives the Professional Development

Traditionally, it was the professional developer who designed the professional development, and this happened outside of the context in which teachers and schools would be participating. However, as professional development has evolved, there has been a move toward having teachers directly drive the professional development. This means it is teachers who decide what they need and how opportunities for their learning should be shaped. The chapters in this volume put forth an alternative to either of these two perspectives on the design of professional development. Within a number of the examples presented, not only is the professional development negotiated between teachers and professional developers, but the authors also argue that this negotiation is necessary in the evolution of the professional development.

The chapter by Hand and her colleagues point out in their case that the teachers were presented options, and wanted to do professional development that connected to their current practice. The teachers participated in choosing the professional development focus they thought would be most helpful. Yet, the professional developers took the teachers' request and considered how to meet the teachers' needs in ways that drew on the professional developers' experience and research-based knowledge, and did so in a way that kept equity central. Battey and Chan described how the professional developers supported teachers to see themselves as math teachers who listened to and built on their students' mathematical thinking and created classrooms that supported learning mathematics with understanding (what the teachers wanted) as the initial step, and who then began to raise the equity issues. Linking this theme to work discussed in Part I, we see that for Rubel, as issues arose in her professional development about teachers and their learning, she was able to adjust to meet the needs of the teachers and still maintain her focus on equity. Edwards discussed how critical the negotiation of professional development is for teacher learning. Edwards expressed a deep respect for teachers as people, what they bring, and what they need, taking what she calls a non-deficit approach to professional development. She argued that we need to understand who teachers are and what they bring, engaging with them in a way that allows teachers and professional development providers to learn together within the context of professional development. Spencer and her colleagues showed how pre-designed professional development, even when adjusted throughout the engagement with teachers, can fail to meet the needs of each of the teachers. In their case, only one teacher was able to bring conceptually challenging mathematics to her students. They believe that at least a part of her success was her disposition toward the students as capable problem solvers.

Conclusions

Spencer and her colleagues remind us that we often don't address equity in professional development work in ways that help teachers reach students in the lowest-performing schools. The fact that many of the authors found ways to take existing approaches to professional development and draw out a more central focus on equity shows that it is possible to make equity central within ongoing professional development work. We were also reminded by Spencer and her colleagues that long-standing and critical constructs within the teaching and learning of mathematics may need to be expanded or changed if we are going to capture what makes a difference for students and support schools and teachers to develop in ways that better meet the needs of low-income students of color in low-performing schools.

Each of the chapters in the book demonstrates the challenge of engaging teachers to consider issues of equity and create more equitable mathematics teaching practice. Fortunately, these chapters also show that we have some theoretically grounded, research-based approaches to do so—approaches with tremendous promise. We were struck by how each of the authors worked to maintain a central focus on equity throughout their work with teachers.

The ideas raised in this volume make a significant contribution to the fields of mathematics education, teacher education, and professional development. The ultimate goal of the collective work is to improve the educational outcomes for students who have traditionally been unsuccessful in mathematics, a group that is disproportionately low-income, nonWhite, and female. We are in awe of the authors' efforts toward this goal and hope their work will enable others to see what is possible.

REFERENCES

Aguirre, J., Celedón-Pattichis, S., Musanti, S., & Anhalt, C. (2010). *K-12 teachers' initial beliefs on teaching mathematics to English learners and Latinas/os*. Manuscript in preparation.
Allen, J., Fabregas, V., Hankins, K. H., Hull, G., Labbo, L., Lawson, H. S., et al. (2002). PhOLKS lore: Learning from photographs, families, and children. *Language Arts, 79*(4), 312–322.
Anyon, J. (1997). *Ghetto schooling: A political economy of urban education reform*. New York: Teachers College Press.
Anzaldua, G. (1987). *Borderlands/la frontera: The new mestiza*. San Francisco: Aunt Lute Book Company.
Apple, M. (1995). Do the standards go far enough? Power, policy and practice in mathematics education. *Journal for Research in Mathematics Education, 23*(5), 412–431.
Ball, D. L. (1996). Teacher learning and the mathematics reforms. *Phi Delta Kappan, 77*(7), 500–515.
Ball, D. L., & Cohen, D. K. (1999). Developing practice, developing practitioners: Toward a practice-based theory of professional education. In L. Darling-Hammond & G. Sykes (Eds.), *Teaching as the learning profession. Handbook of policy and practice* (pp. 3–31). San Francisco: Jossey-Bass.
Ball, D. L., Lubienski, S. T., & Mewborn, D. (2001). Research on teaching mathematics: The unresolved problem of teachers' mathematical knowledge. In V. Richardson (Ed.), *Handbook of research on teaching* (pp. 433–456). Washington, D.C.: American Educational Research Association.
Ball, D. L., Thames, M. H., & Phelps, G. (2008). Content knowledge for teaching: What makes it special? *Journal of Teacher Education, 59*(5), 389–407.
Banks, J. A. (2004). Multicultural education: Historical development, dimensions, and practice. In J. A. Banks & C. A. M. Banks (Eds.), *Handbook of research on multicultural education* (2nd ed., pp. 3–29). San Francisco: Jossey-Bass.
Barron, B. (2003). When smart groups fail. *Journal of the Learning Sciences, 12*(3), 307–359.
Bartolomé, L., & Trueba, E. (2000). Beyond the politics of schools and the rhetoric of fashionable pedagogies: The significance of teacher ideology. In E. B. Trueba & L. Bartolomé (Eds.), *Immigrant voices: In search of educational equity* (pp. 277–292). New York: Rowman & Littlefield.
Battey, D. (2004). *Designing an approach to assess content-specific teacher knowledge*. Unpublished doctoral dissertation. University of California, Los Angeles.
Battey, D., Foote, M. Q., Spencer, J., Taylor, E. V., & Wager, A. (2007). *Professional development at the intersection of mathematics and equity*. Paper presented at the Research Pre-session of the Annual Meeting of the National Council of Teachers of Mathematics. Atlanta, GA.

Battey, D., & Franke, M. L. (2008). Transforming identities: Understanding teachers across professional development and classroom practice. *Teacher Education Quarterly, 35*(3), 127–149.
Battey, D., & Franke, M. L. (2010). Integrating professional development on mathematics and equity: Making the case for challenging the metanarrative. Manuscript submitted for publication.
Bell, L. A., Washington, S., Weinstein, G.,& Love, B. (1997). Knowing ourselves as instructors. In M. Adams, L.A. Bell & P. Griffin (Eds.), *Teaching for diversity and social justice: A sourcebook* (pp. 299–310). New York: Routledge.
Bielenberg, B., & Wong Fillmore, L. (2004/2005). The English they need for the test. *Educational Leadership, 64*(4), 45–49.
Biggs, S. F., Rosman, A. J., & Sergenian, G. K. (1993). Methodological issues in judgment and decision-making research: Concurrent verbal protocol validity and simultaneous traces of process. *Journal of Behavioral Decision Making, 6,* 187–206.
Birky, G., Chazan, D., & Farlow, K. (2008). *Deliberately departing from the curriculum guide in search of coherence and meaning: Madison Morgan's mathematics instruction in an urban school.* Paper presented at the Annual Meeting of the American Education Research Association. New York.
Boaler, J. (2006a). How a detracked mathematics approach promoted respect, responsibility, and high achievement. *Theory into Practice, 45*(1), 40–46.
Boaler, J. (2006b). Promoting respectful learning. *Educational Leadership, 63*(5), 74–78.
Boaler, J., & Greeno, J. (2000). Identity, agency, and knowing in mathematical worlds. In J. Boaler (Ed.), *Multiple perspectives on mathematics teaching and learning* (pp. 171–200). Westport, CT: Ablex.
Boaler, J., & Staples, M. (2008). Creating mathematical futures through an equitable teaching approach: The case of Railside School. *Teachers College Record, 110*(3), 608–645.
Boreham, N., & Morgan, C. (2004). A sociocultural analysis of organisational learning. *Oxford Review of Education, 30,* 307–325.
Braddock III, J. H., & Dawkins, M. P. (1993). Ability grouping, aspirations, and attainments: Evidence from the National Educational Longitudinal Study of 1988. *The Journal of Negro Education, 62*(3), 342–336.
Bransford, J., Brown, A., & Cocking, R. (Eds.). (2000). *How people learn: Brain, mind, experience, and school.* Expanded Edition. Washington, D.C.: National Academies Press.
Butcher, E., & Scofield, M. E. (1984). The use of a standardized simulation and process tracing for studying clinical problem-solving competence. *Counselor Education and Supervision, 24*(1), 70–85.
Cabello, B., & Burstein, N. D. (1995). Examining teachers' beliefs about teaching in culturally diverse classrooms. *Journal of Teacher Education, 46*(4), 285–294.
Carpenter, T. P., Blanton, M. L., Cobb, P., Franke, M. L., Kaput, J., & McClain, K. (2004). *Scaling up innovative practices in mathematics and science.* Madison, WI: National Center for Improving Student Learning and Achievement in Mathematics and Science.
Carpenter, T. P., Fennema, E., & Franke, M. L. (1996). Cognitively guided instruction: A knowledge base for reform in primary mathematics instruction. *The Elementary School Journal, 97*(1), 3–20.
Carpenter, T. P., Fennema, E., Franke, M. L., Levi, L., & Empson, S. (1999). *Children's mathematics: Cognitively guided instruction.* Portsmouth, NH: Heinemann.
Carpenter, T. P., Fennema, E., Peterson, P., Chi-Pang, C., & Loef, M. (1989). Using knowledge of children's mathematics thinking in classroom teaching: An experimental study. *American Educational Research Journal, 24*(4), 499–531.
Celedón-Pattichis, S. (2008). "What does that mean?": Drawing on Latino and Latina students' language and culture to make mathematical meaning. In M. W. Ellis (Ed.), *Mathematics for*

References

every student: Responding to diversity, grades 6–8 (pp. 59–73). Reston, VA: National Council of Teachers of Mathematics.

Centinela Valley Union High School District. (2006). *Leuzinger High School: School accountability report card, 2005–2006.* San Francisco: School Wise Press.

Chu, H., & Rubel, L. (2009). Learning to teach mathematics in urban high schools: Unearthing the layers. Manuscript submitted for publication

Chubbuck, S.M., & Zembylas, M. (2008). The emotional ambivalence of socially just teaching: A case study of a novice urban school teacher. *American Educational Research Journal, 45*(2), 274–318.

Civil, M. (1995a). *Bringing the mathematics to the foreground.* Paper presented at the American Educational Research Association Annual Meeting, San Francisco.

Civil, M. (1995b). *Everyday mathematics, "mathematician" mathematics," and school mathematics: Can we (and should we) bring these three cultures together?* Paper presented at the Annual Meeting of the American Educational Research Association, San Francisco.

Civil, M. (1998). *Linking home and school: In pursuit of a two-way mathematical dialogue.* Paper presented at the 22nd Conference of the International Group for the Psychology of Mathematics Education, Stellenbosch, South Africa.

Civil, M. (2002). Everyday mathematics, mathematicians' mathematics, and school mathematics: Can we bring them together? In M. Brenner & J. Moschkovich (Eds.), *Everyday and academic mathematics in the classroom* (Vol. 11, pp. 40–62). Reston, VA: Journal for Research in Mathematics Education Monograph.

Civil, M. (2006). Building on community knowledge: An avenue to equity in mathematics education. In N. Nasir and P. Cobb (Eds.), *Improving access to mathematics: Diversity and equity in the classroom* (pp. 105–117). New York: Teachers College Press.

Civil, M., & Andrade, R. (2002). Transitions between home and school mathematics: Rays of hope amidst the passing clouds. In G. de Abreu, A.J. Bishop, & N.C. Presmeg (Eds.), *Transitions between contexts of mathematical practices* (pp. 149–169). Boston: Kluwer.

Cobb, P. (2000). Conducting classroom teaching experiments in collaboration with teachers. In A. Kelly, & R. Lesh (Eds.), *Handbook of research design in mathematics and science education* (pp. 307–334). Mahwah, NJ: Erlbaum.

Cochran-Smith, M. (1999). Learning to teach for social justice. In G. Griffin (Ed.), *98th yearbook of NSSE: Teacher education for a new century: Emerging perspectives, promising practices, and future possibilities* (pp. 114–145). Chicago: University of Chicago Press.

Cochran-Smith, M., & Lytle, S. (1999). Relationships of knowledge and practice: Teacher learning in communities. *Review of Research in Education, 24,* 249–305.

Cohen, E. G. (1994). Restructuring the classroom: Conditions for productive small groups. *Review of Educational Research, 64*(1), 1–35.

Cohen, E. G., & Goodlad, J. I. (1994). *Designing groupwork: Strategies for the heterogeneous classroom* (2nd ed.). New York: Teachers College Press.

Cohen, E. G., & Lotan, R. A. (1997a). *Working for equity in heterogeneous classrooms.* New York: Teachers College Press.

Cohen, E. G., & Lotan, R. A. (Eds.). (1997b). *Working for equity in heterogeneous classrooms: Sociological theory in practice.* New York: Teachers College Press.

Cole, M. (1996). *Cultural psychology.* Cambridge, MA: Harvard University Press.

Confrey, J., & Lachance, A. (2000). A research design model for conjecture-driven teaching experiments. In A. Kelly, & R. Lesh (Eds.), *Handbook of research design in mathematics and science education* (pp. 231–266). Mahwah, NJ: Elrbaum.

Connelly, F., & Clandinin, D. (1990). Stories of experience and narrative inquiry. *Educational Researcher, 19*(2), 3–14.

Costa, A., & Garmston, R. (2002). *Cognitive coaching: A foundation for renaissance schools.* Norwood, MA: Christopher-Gordon.

Covey, J. A., & Lovie, A. D. (1998). Information selection and utilization in hypothesis testing: A comparison of process-tracing and structural analysis techniques. *Organizational Behavior and Human Decision Processes, 75,* 56–74.

Crockett, M. D. (2002). Inquiry as professional development: Creating dilemmas through teachers' work. *Teaching and Teacher Education, 18,* 609–624.

Cummins, J. (1986). The role of primary language development in promoting educational success for language minority students. In California State Department of Education (Ed.), *Schooling and language minority students: A theoretical framework* (pp. 3–50). Los Angeles: Evaluation, Dissemination, and Assessment Center.

Cummins, J. (2001). Empowering minority students: A framework for intervention. *Harvard Educational Review, 71*(4), 649–675.

Darling-Hammond, L. (1997). *The right to learn: A blueprint for creating schools that work.* San Francisco: Jossey-Bass.

Darling-Hammond, L. (2002). Educating a profession for equitable practice. In L. Darling-Hammond, J. French & S. P. Garcia-Lopez (Eds.), *Learning to teach for social justice* (pp. 201–212). New York: Teachers College Press.

Decorte, E. (1995). Fostering cognitive growth: A perspective from research on mathematics learning and instruction. *Educational Psychologist, 30,* (1), 37–46.

Delgado-Gaitan, M. (1987). Traditions and transitions in the learning process of Mexican children: An ethnographic view. In G. Spindler & L. Spindler (Eds.), *Interpretive ethnography of education: At home and abroad* (pp. 333–359). Hillsdale, NJ: Erlbaum.

Delgado, R., & Stefancic, J. (1992). Images of the outsider in American law and culture: Can free expression remedy systemic ills? *Cornell Law Review, 77,* 1258–1297.

DiME. (2005). *Why they fail: Unpacking everyday explanations of the achievement gap within research on differential mathematics achievement.* Symposium conducted at the Annual Meeting of the American Educational Research Association. Montreal.

DiME. (2007). Culture, race, power, and mathematics education. In F. Lester (Ed.), *Second handbook of research on mathematics teaching and learning: A project of the National Council of Teachers of Mathematics* (pp. 405–433). Charlotte, NC: Information Age.

Drake, C., & Sherin, M. G. (2006). Practicing change: Curriculum adaptation and teacher narrative in the context of mathematics education reform. *Curriculum Inquiry, 36*(2), 153–187.

Dunn, T. (2005). Engaging prospective teachers in critical reflection: Facilitating a disposition to teach mathematics for diversity. In A. Rodriguez, & R.S. Kitchen (Eds.), *Preparing mathematics and science teachers for diverse classrooms.* Mahwah, NJ: Erlbaum.

Egi, T. (2008). Investigating stimulated recall as a cognitive measure: Reactivity and verbal reports in SLA research methodology. *Language Awareness, 17* (3), 212–228.

Ehrenreich, B. (2001). *Nickel and dimed: On (not) getting by in America.* New York: Henry Holt.

Eisenhart, M., Borko, H., Underhill, R., Brown, C., Jones, D., & Agard, P. (1993). Conceptual knowledge falls through the cracks: Complexities of learning to teach mathematics for understanding. *Journal for Research in Mathematics Education, 24*(1), 8–40.

Elliot, R., Lesseig, K., Kazemi, E., & Kelly-Peterson, M. (2009). *Sociomathematical norms for explanation in professional development: Opportunities for teacher leaders to learn mathematical content for teaching.* Paper presented at the Annual Meeting of the American Educational Research Association. San Diego, CA.

Erickson, F. (1986). Qualitative methods in research on teaching. In M. C. Wittrock (Ed.), *Handbook of research on teaching* (3rd ed. pp. 119–161). New York: Macmillan.

REFERENCES

Erlandson, D., Harris, E., Skipper, B., & Allen, S. (1993). *Doing naturalistic inquiry: A guide to methods.* Newbury Park, CA: Sage.

Esmonde, I. (2009). Ideas and identities: Supporting equity in cooperative mathematics learning. *Review of Educational Research, 79*(2), 1008–1043.

Eubanks, E., Parish, R., & Smith, D. (1997). Changing the discourse in schools. In P. M. Hall (Ed.), *Race, ethnicity and multiculturalism.* New York: Garland.

Fendel, D., Resek, D., Alper, L., & Fraser, S. (1998). *Interactive mathematics program: Year 2.* Emeryville, CA: Key Curriculum Press.

Fennema, E., Carpenter, T. P., Franke, M. L., Levi, L., Jacobs, V. R., & Empson, S. B. (1996). A longitudinal study of learning to use children's mathematical thinking in mathematics instruction. *Journal for Research in Mathematics Education, 27*(4), 403–434.

Fennema, E., & Franke, M. (1992). Teachers' knowledge and its impact. In D. Grouws (Ed.), *Handbook of research on mathematics teaching and learning* (pp. 147–164). New York: Macmillan.

Fennema, E., & Sherman, J. A. (1976). Fennema-Sherman mathematics attitude scales: Instruments designed to measure attitudes toward the learning of mathematics by females and males. *JSAS: Catalog of Selected Documents in Psychology, 6*(2) 31.

Fernandez, C., Cannon, J., & Chokshi, S. (2003). A US-Japan lesson study collaboration reveals critical lenses for examining practice. *Teaching and Teacher Education, 19*(2), 171–185.

Fernandez, C., & Yoshida, M. (2000). *Lesson study as a model for improving teaching: Insights, challenges and a vision for the future.* Paper presented at the Wingspread Conference, In the Eye of the Storm: Improving Teaching Practices to Achieve Higher Standards. Racine, WI.

Flores, A. (2007). Examining disparities in mathematics education: Achievement gap or opportunity gap? *The High School Journal, 91*(1), 29–42.

Foote, M. Q. (2006). *Supporting teachers in situating children's mathematical thinking within their lived experience.* Unpublished doctoral dissertation, University of Wisconsin, Madison.

Foote, M. Q. (2008). Addressing the needs of struggling learners. *Teaching Children Mathematics, 14*(6), 340–342.

Foote, M. Q. (2009). Stepping out of the classroom: Building teacher knowledge for developing classroom practice. *Teacher Education Quarterly, 36*(3), 39–53.

Foote, M. Q., Bartell, T., & Wager, A. (2007). *Looking outside the mathematics classroom: Professional development that integrates mathematics and lived experiences.* Paper presented at the North American Chapter of the International Group for the Psychology of Mathematics Education. Lake Tahoe, NV.

Foster, H. L. (1986). *Ribin', jivin' and playin' the dozens.* Cambridge, MA: Ballinger.

Franke, M. L. (2009). *Identity, equity and professional development.* Plenary talk given at the Research Pre-session of the Annual Meeting of the National Council of Teachers of Mathematics. Washington, D.C.

Franke, M. L., Carpenter, T. P., Levi, L., & Fennema, E. (2001). Capturing teachers' generative change: A follow-up study of professional development in mathematics. *American Education Research Journal, 38* (3), 653–691.

Franke, M.L., & Chan, A. (2009). A practice-based approach to school-based professional development in elementary mathematics. Manuscript submitted for publication.

Franke, M. L., & Kazemi, E. (2001). Learning to teach mathematics: Focus on student thinking. *Theory into Practice, 40*(2), 102–109.

Franke, M. L.,, & Kazemi, E. (2001). Teaching as learning within a community of practice: Characterizing generative growth. In T. Woods, B. Nelson, & J. Warfield (Eds.), *Beyond classical pedagogy in elementary mathematics: The nature of facilitative teaching* (pp. 47–74). Mahwah, NJ: Erlbaum.

Franke, M. L., Kazemi, E., & Battey, D. (2007). Mathematics teaching and classroom practice. In F. Lester (Ed.), *Second handbook of research on mathematics teaching and learning: A project of the National Council of Teachers of Mathematics* (pp. 225–256). Charlotte, NC: Information Age.

Franke, M. L., Kazemi, E., Shih, J., Biagetti, S., & Battey, D. (2005). Work in mathematics: One school's journey. In T. A. Romberg, T. P. Carpenter, & F. Dremock (Eds.) *Understanding mathematics and science matters* (pp. 209–230). Mahwah, NJ: Erlbaum.

Frankenstein, M. (1995). Equity in mathematics education: Class in a world outside of class. In W. G. Secada, E. Fennema, & L. B. Adajian (Eds.), *New directions for equity in mathematics education* (pp. 165–190). New York: Cambridge University Press.

Freire, P. (1983). The banking concept of education. In H. Giroux, & D. Purpel (Eds.) *The hidden curriculum and moral education*. Berkeley, CA: McCutchan.

Freire, P. (1993). *Pedagogy of the oppressed*. New York: Continuum.

Gau, T. (2005). *Learning to teach mathematics for social justice*. Unpublished Doctoral Dissertation, University of Wisconsin, Madison. Madison, WI.

Gay, G. (2002). Preparing for culturally responsive teaching. *Journal of Teacher Education, 53*(2), 106–116.

Gee, J. (1989). Literacy, discourse, and linguistics: Introduction. *Journal of Education, 171*(1), 5–17.

Gibbons, P. (2009). *English learners, academic literacy, and thinking: Learning in the challenge zone*. Portsmouth, NH: Heinemann.

Giroux, H., Lankshear, C., McLaren, P., & Peters, M. (1996). *Counternarratives: Cultural studies and critical pedagogies in postmodern spaces*. New York: Routledge.

Glaser, B. G., & Strauss, A. L. (1967). *The discovery of grounded theory: Strategies for qualitative research*. Chicago: Aldine.

González, N., Andrade, R., Civil, M., & Moll, L. (2001). Bridging funds of distributed knowledge: Creating zones of practices in mathematics. *Journal of Education for Students Placed at Risk, 6*(1&2), 115–132.

González, N., Moll, L., & Amanti, C. (2005). *Funds of knowledge: Theorizing practices in households, communities, and classrooms*. Mahwah, NJ: Erlbaum.

Griffin, P. (1997). Facilitating social justice education courses. In M. Adams, L. A. Bell & P. Griffin (Eds.), *Teaching for diversity and social justice: A sourcebook* (pp. 279–298). New York: Routledge.

Grossman, P., & McDonald, M. (2008). Back to the future: Directions for research in teaching and teacher education. *American Educational Research Journal, 45*(1), 184–205.

Grossman, P., Wineburg, S., & Woolworth, S. (2001). Toward a theory of community. *Teachers College Record, 103*, 942–1012.

Guberman, S. R. (2004). A comparative study of children's out-of-school activities and arithmetical achievements. *Journal for Research in Mathematics Education, 35*(2). p. 117–150.

Gutiérrez, K., & Rogoff, B. (2003). Cultural ways of learning: Individual traits or repertoires of practice. *Educational Researcher, 32*(5), 19–25.

Gutiérrez, R. (2002). Enabling the practice of mathematics teachers in context: Toward a new equity research agenda. *Mathematical Thinking and Learning, 4*(2/3), 145–187.

Gutstein, E. (2003). Teaching and learning mathematics for social justice in an urban, Latino school. *Journal for Research in Mathematics Education, 34*(1), 37–73.

Gutstein, E. (2006). *Reading and writing the world with mathematics: Toward a pedagogy for social justice*. New York: Routledge.

Gutstein, E., Lipman, P., Hernandez, P., & de los Reyes, R. (1997). Culturally relevant mathematics teaching in a Mexican American context. *Journal for Research in Mathematics Education, 28*(6), 709–737.

References

Gutstein, E., & Peterson, B. (2005). *Rethinking mathematics: teaching social justice by the numbers.* Milwaukee, WI: Rethinking Schools.

Hand, V. (2003). Reframing participation: Meaningful mathematical activity in diverse classrooms. Unpublished doctoral dissertation, Stanford University.

Hart, L. (2002). Preservice teachers' beliefs and practice after participating in an integrated content/methods course. *School Science and Mathematics, 102,* 4–14.

Hatfield, M., & Bitter, G. (1994). A multimedia approach to the professional development of teachers: A virtual classroom. In D. B. Aichele (Ed.), *NCTM 1994 yearbook: Professional development for teachers of mathematics* (pp. 102–115). Reston, VA: National Council of Teachers of Mathematics.

Heaton, R.T. (1992). Who is minding the mathematics content?: A case study of a fifth-grade teacher. *The Elementary School Journal, 93*(2), 153–162.

Henderson, L., & Tallman, J. (1998). Teaching effectively with electronic databases: Paradigms suggested by interactive changes in teachers' mental model. *Proceedings of the World Conference on Educational Multimedia and Hypermedia and World Conference on Educational Telecommunications,* Freiburg, Germany, 143–150.

Hiebert, J., & Carpenter, T. P. (1992). Learning and teaching with understanding. In D. Grouws (Ed.), *Handbook of research on mathematics teaching and learning* (pp. 65–97). NY: Macmillan.

Hiebert, J., Gallimore, R., Garnier, H., Givvin, K. B., Hollingsworth, H., Jacobs, J., Chui, A. M., Wearne, D., Smith, M., Kersting, N., Manaster, A., Tseng, E., Etterbeek, W., Manaster, C., Gonzales, P., & Stigler, J. (2003). *Teaching mathematics in seven countries: Results from the TIMSS 1999 Video Study,* NCES (2003–013), U.S. Department of Education. Washington, D.C.: National Center for Education Statistics.

Hill, H., Rowan, B., & Ball, D.L. (2005). Effects of teachers' mathematical knowledge for teaching on student achievement. *American Educational Research Journal, 42*(2), 371–406.

Hilliard, A. (1974). Restructuring teacher education for multicultural imperatives. In W.A. Hunter (Ed.), *Multicultural education through competency-based teacher education.* Washington, D.C.: American Association of Colleges of Teacher Education.

Hilliard, A. (1989). Teachers and cultural styles in pluralistic society. *NEA Today,7*(6), 65–69.

Himley, M., & Carini, P. (Eds.). (2000). *From another angle: Children's strengths and school standards* New York: Teachers College Press.

Ho, K.M. (2009). *Race and Equity in the Math Class: Teacher Learning via Artifacts.* Unpublished doctoral dissertation, University of California, Los Angeles.

Holland, D., Lachicotte, W., Skinner, D., & Cain, C. (2001). *Identity and agency in cultural worlds* Cambridge, MA: Harvard University Press.

Horn, I.S. (2007). Fast kids, slow kids, lazy kids: Framing the mismatch problem in mathematics teachers' conversations. *Journal of the Learning Sciences, 16(1),* 37–79.

Howard, G. (1999). *We can't teach what we don't know: White teachers, multiracial schools.* New York: Teachers College Press.

Howard, T. C. (2003). Culturally relevant pedagogy: Ingredients for critical teacher reflection. *Theory into Practice, 42*(3), 195–202.

Hufferd-Ackles, K., Fuson, K.C., & Sherin, M.G. (2004). Describing levels and components of a math-talk learning community. *Journal for Research in Mathematics Education, 35*(2), 81–116.

Jacobs, V. R., Franke, M. L., Carpenter, T. P., Levi, L., & Battey, D. (2007). Professional development focused on children's algebraic reasoning in elementary school. *Journal for Research in Mathematics Education, 38*(3), 258–288.

Jamar, I., & Pitts, V.R. (2005). High expectations: A "How" of achieving equitable mathematics classrooms, *Negro Educational Review, 56* (2&3), 127–134.

Jensen, B. T. (2007). The relationship between Spanish use in the classroom and the mathematics achievement of Spanish-speaking kindergartners. *Journal of Latinos and Education, 6*(3), 267–280.

Johnston, L. D., O'Malley, P. M., Bachman, J. G., & Schulenberg, J. E. (2003). Monitoring the future: National survey results on drug use, 1975–2003. Betheseda, MD: National Institute of Drug Abuse.

Kazemi, E. (2008). School development as a means of improving mathematics teaching and learning: Towards multidirectional analyses of learning across contexts. In K. Krainer & T. Wood (Eds.), *Participants in mathematics teacher education: Individuals, teams, communities and networks* (pp. 209–230). Rotterdam, Netherlands: Sense.

Kazemi, E., & Franke, M. L. (2004). Teacher learning in mathematics: Using student work to promote collective inquiry. *Journal of Mathematics Teacher Education, 7,* 203–235.

Khisty, L. L. (1995). Making inequality: Issues of language and meanings in mathematics teaching with Hispanic students. In W. G. Secada, E. Fennema, & L.B. Adajian (Eds.), *New directions for equity in mathematics education* (pp. 279–298). New York: Cambridge University Press.

Knapp, M. S. (1995). *Teaching for meaning in high poverty classrooms.* New York: Teachers College Press.

Koehn, C. (2009). Math is more than numbers: A model for forging connections between equity, teacher participation, and professional development. Unpublished doctoral dissertation. University of California, Los Angeles.

Kohler, A., & Lazarín, M. (2007). *Hispanic education in the United States: Statistical brief No. 8.* Washington, D.C.: National Council of La Raza.

Kozol, J. (2006). *The shame of a nation: The restoration of apartheid schooling in America.* New York: Crown.

Ladson-Billings, G. (1994a). *The dreamkeepers: Successful teachers of African American children.* San Francisco: Jossey-Bass.

Ladson-Billings, G. (1994b). Who will teach our children?: Preparing teachers to successfully teach African American students. In E. R. Hollins, J. E. King, & W. C. Hayman (Eds.), *Teaching diverse populations: Formulating a knowledge base* (pp. 129–142). Albany, NY: State University of New York Press.

Ladson-Billings, G. (1995). Toward a theory of culturally relevant pedagogy. *American Educational Research Journal, 32*(3), 465–491.

Ladson-Billings, G. (1997). It doesn't add up: African American students' mathematics achievement. *Journal for Research in Mathematics Education, 28*(6), 697–708.

Ladson-Billings, G. (1998). Just what is critical race theory and what's it doing in a nice field like education? *International Journal of Qualitative Studies in Education, 11*(1), 7–24.

Ladson-Billings, G. (2001). *Crossing over to Canaan.* San Francisco: Jossey-Bass.

Ladson-Billings, G., & Tate, W. F. (1995). Toward a critical race theory of education. *Teachers College Record, 97*(1), 47–68.

Lampert, M. (2003). *Teaching problems and the problems of teaching.* New Haven, CT: Yale University Press.

Lappan, G. (1997). The challenges of implementation: Supporting teachers. *American Journal of Education, 106,* 207–239.

Lappan, G. (1998). Pedagogical implication for problem-centered teaching. In Mathematics Sciences Education Board, (Ed.), *High school mathematics at work: Essays and examples for the education of all students* (pp. 132–140). Washington, D.C.: National Academy Press.

Lave, J. (1991). Situated learning in communities of practice. In L. B. Resnick, J. M. Levine, & S. D. Teasley (Eds.), *Perspectives on socially shared cognition* (pp. 63–82). Washington, D.C.: American Psychological Association.

References

Lave, J. (1996). Teaching as learning in practice. *Mind, Culture, and Activity, 3*, 149–164.

Lave, J., & Wenger, E. (1991). *Situated learning: legitimate peripheral participation*. Cambridge, UK: Cambridge University Press.

Lewis, C., Perry, R., & Hurd, J. (2004). A deeper look at lesson study. *Educational Leadership, 61*(5), 18–22.

Lewis, C., & Tsuchida, I. (1998). A lesson is like a swiftly flowing river: How research lessons improve Japanese education. *American Educator, 22*(4), 12–17.

Lincoln, Y. S., & Guba, E. G. (1985). *Naturalistic inquiry*. Newbury Park, CA: Sage.

Lindholm-Leary, J. K. (2001). *Dual language education*. Bristol, UK: Multilingual Matters.

Little, J. W. (1993). Teachers' professional development in a climate of educational reform. *Educational Evaluation and Policy Analysis, 15*(2), 129–151.

Little, J. W. (2002). Locating learning in teachers' community of practice: Opening up problems of analysis in records of everyday work. *Teaching and Teacher Education, 18,* 917–946.

Little, J. W. (2005). Looking at student work in the United States: A case of competing impulses in professional development. In C. Day & J. Sachs (Eds.), *International handbook on the continuing professional development of teachers* (pp. 94–118). Columbus, OH: Open University Press.

López-Bonilla, G. (2002). Los programas de inmersión bilingüe y la adquisición del discurso académico. *Bilingual Research Journal, 26*(3), 525–536.

Lotan, R. A. (2003). Group-worthy tasks. *Educational Leadership, 60*(7), 72–75.

Lubienski, S. T. (2008). On "gap gazing" in mathematics education: The need for gaps analyses. *Journal for Research in Mathematics Education, 39*(4), 350–356.

Ma, L. (1999). *Knowing and teaching elementary mathematics*. Mahwah, NJ: Erlbaum.

Mao, J., & Benbasat, I. (1998). Contextual access to knowledge: Theoretical perspectives and a process-tracing study. *Information Systems Journal, 8,* 217–239.

Martin, D. B. (2000). *Mathematics success and failure among African American youth*. Mahwah, NJ: Erlbaum.

Martin, D. B. (2003). Hidden assumptions and unaddressed questions in mathematics for all rhetoric. *The Mathematics Educator. 13*(2), 7–21.

Martin, D. B. (2007). Mathematics learning and participation in the African American context: The co-construction of identity in two intersecting realms of experience. In N. Nasir & P. Cobb (Eds.), *Improving access to mathematics: Diversity and equity in the classroom* (pp. 146–158). New York: Teachers College Press.

Martin, R. J. & Van Gunten, D. M. (2002). Reflected identities: Applying positionality and multicultural reconstructionism in teacher education. *Journal of Teacher Education, 53*(1), 44–54

Maryland, P., Patching, W., & Putt, I. (1992). Thinking while studying: A process tracing study of distance learners. *Distance Education, 13*(2), 193–217.

McLaughlin, M., & Talbert, J. (2001). *Professional communities and the work of high school teaching* Chicago: University of Chicago Press.

McLeod, D. (1994). Research on affect and mathematics learning in the JRME: 1970 to the present. *Journal for Research in Mathematics Education. 25*(6), 637–47.

Merriam, S. B. (1998). *Qualitative research and case study applications in education* (2nd Ed.). San Francisco: Jossey-Bass.

Moje, E. (2007). Developing socially just subject-matter instruction: A review of the literature on disciplinary literacy teaching. *Review of Research in Education, 31,* 1–44.

Moll, L. C. (1992). Bilingual classroom studies and community analysis: Some recent trends. *Educational Researcher, 21*(2), 20–24.

Moll, L. C., Amanti, C., Neff, D., & Gonzalez, N. (1992). Funds of knowledge for teaching: A qualitative approach to connect homes and classrooms. *Theory into Practice, 31*(1), 132–141.

Moll, L. C., & Greenberg, J. (1990). Creating zones of possibilities: Combining social contexts for instruction. In L. C. Moll (Ed.), *Vygotsky and education* (pp. 319–348). Cambridge, UK: Cambridge University Press.

Moschkovich, J. (2007). Bilingual mathematics learners: How views of language, bilingual learners, and mathematical communication impact instruction. In N. Nassir & P. Cobb (Eds.), *Diversity, equity, and access to mathematical ideas* (pp. 121–144). New York: Teachers College Press.

Moses, R.P., & Cobb, C.E. (2001). *Radical equations: Math literacy and civil rights*. Boston: Beacon Press.

Mullens, J., Murnane, R., & Willet, J. (1996). The contribution of training and subject matter knowledge of teaching effectiveness: A multilevel analysis of longitudinal evidence from Belize. *Comparative Education Review, 40*(2), 139–157.

Musanti, S. I., Celedón-Pattichis, S., & Marshall, M. E. (2009). Reflections on language and mathematics problem solving: A case study of a bilingual first grade teacher. *Bilingual Research Journal, 32*(1), 25–41.

National Commission on Excellence in Education (1983). *A Nation at risk: The imperative for educational reform*. Washington, D.C.: U.S. Department of Education.

National Council of Teachers of Mathematics. (1989). *Curriculum and evaluation standards for school mathematics*. Reston, VA: Author.

National Council of Teachers of Mathematics (1991). *Professional standards for teaching mathematics*. Reston, VA: Author.

National Council of Teachers of Mathematics. (2000). *Principles and standards for school mathematics*. Reston, VA: Author.

National Council of Teachers of Mathematics. (2008). *Position statement on teaching mathematics to English language learners*. Retrieved on October 30, 2008 from http://www.nctm.org/about/content.aspx?id=16135

National Research Council. (1989). *Everybody counts: A report to the nation on the future of mathematics education*. Washington, D.C.: National Academy Press.

National Research Council (1990). *Reshaping school mathematics: A philosophy and framework for curriculum*. Washington, D.C. National Academy Press.

National Research Council. (2001). *Adding it up: Helping children learn mathematics*. Washington, D.C.: National Academy Press.

Nieto, S. (2004). *Affirming diversity* (4th ed.). Boston: Pearson.

North, C. E. (2006). More than words? Delving into the substantive meaning(s) of "social justice" in education. *Review of Educational Research, 76*(4), 507–535.

North, C. E. (2008). What is all this talk about 'social justice'?: Mapping the terrain of education's latest catchphrase. *Teachers College Record, 110*(6), 1182–1206.

Oakes, J. (1985). *Keeping tracks: How schools structure inequality*. New Haven: Yale University Press.

Oakes, J. (1990). *Multiplying inequalities: The effects of race, social class, and tracking on opportunities to learn math and science*. Santa Monica, CA: RAND.

Oakes, J. (2005). *Keeping tracks: How schools structure inequality* (2nd ed.). New Haven: Yale University Press.

Oakes, J., Joseph, R., & Muir, K. (2004). Access and achievement in mathematics and science: Inequalities that endure and change. In J. A. Banks and C. A. M. Banks (Eds.), *The handbook of multicultural education* (2nd ed., pp. 69–90). San Francisco: Jossey-Bass.

Park, J. (2008). *Preparing secondary mathematics teachers in urban schools: Using video to develop understanding on equity-based discourse and classroom norms*. Unpublished doctoral dissertation, University of California, Los Angeles.

REFERENCES

Patrick, J., & James, N. (2004). Process tracing of complex cognitive work tasks. *Journal of Occupational and Organizational Psychology, 77,* 259–280.

Philipp, R. A. (2007). Mathematics teachers' beliefs and affect. In F. K. Lester (Ed.), *Second handbook of research on mathematics teaching and learning: A project of the National Council of Teachers of Mathematics* (pp. 257–315). Charlotte, NC: Information Age.

Pollock, M. (2004). *Colormute: Race talk dilemmas in an American school.* Princeton, NJ: Princeton University Press.

Putnam, R. (1992). Teaching the "hows" of mathematics for everyday life: A case study of a fifth-grade teacher. *The Elementary School Journal, 93*(2), 163–177.

Putnam, R., & Borko, H. (2000). What do new views of knowledge and thinking have to say about research on teacher learning? *Educational Researcher, 29*(1), 4–15.

Reed, J. (2006). *Secondary mathematics teachers' adjustment to teaching practice in tracked classrooms* Unpublished doctoral dissertation. University of Georgia, Athens, GA.

Rist, R., (1970). Student social class and teacher expectations. *Harvard Educational Review 40,* 411–451.

Rodriguez, A. J., & Kitchen, R. S. (Eds.). (2005). *Preparing mathematics and science teachers for diverse classrooms.* Mahwah, NJ: Lawrence Erlbaum Associates.

Rogoff, B. (1997). Evaluating development in the process of participation: Theory, methods, and practice building on each other. In E. Amsel & K. A. Renninger (Eds.), *Change and development: Issues of theory, method, and application.* Mahwah, NJ: Erlbaum.

Ron, P. (1999). Spanish-English language issues in the mathematics classroom. In L. Ortiz-Franco, N. G. Hernandez, & Y. de la Cruz (Eds.), *Changing the faces of mathematics: Perspectives on Latinos* (pp. 23–33). Reston, VA: NCTM.

Rousseau, C., & Tate, W. F. (2003). No time like the present: Reflecting on equity in school mathematics. *Theory into Practice, 42*(3), 210–216.

Ryu, A. (2006). *A study of teacher learning and professional development through collaborative reflection on artifacts of practice.* Unpublished doctoral dissertation, University of California, Berkeley.

Sallee, T., Kysh, J., Kasaimatis, E., & Hoey, B. (2002*). College Preparatory Mathematics.* Davis, CA: CPM Educational Program.

Santagata, R., Kersting, N., Givvin, K., & Stigler, J.W. (2009). *Problem implementation as a lever for change: An experimental study of the effects of a professional development program on students' mathematics learning.* Manuscript submitted for publication.

Secada, W. G. (1989). Agenda setting, enlightened self-interest, and equity in mathematics education. *Peabody Journal of Education, 66*(2), 22–56.

Secada, W. G. (1992). Race, ethnicity, social class, language, and achievement in mathematics. In D. Grows (Ed.), *Handbook of research on mathematics teaching and learning* (pp. 623–660). New York: Macmillan.

Secada, W. G. (1995). Social and critical dimensions for equity in mathematics education. In W. Secada, E. Fennema, & L. B. Adajian (Eds.), *New directions for equity in mathematics education* (pp. 146–164). New York: Cambridge University Press.

Secada, W. G., & de la Cruz, Y. (1996). Teaching mathematics for understanding to bilingual students. In J. L. Flores (Ed.), *Children of la frontera: Bilingual efforts to serve Mexican migrant and immigrant students.* (ERIC Document Reproduction Service No. ED393646).

Sherin, M. G., & van Es, E. (2005). Using video to support teachers' ability to notice classroom interactions. *Journal of Technology and Teacher Education, 13*(3), 475–491.

Shulman, L.S. (1987). Knowledge and teaching: Foundations of the new reform. *Harvard Educational Review, 57* (1), 1–22.

Singleton, G. E., & Linton, C. (2007). *Courageous conversations about race: A field guide for achieving equity in schools.* Thousand Oaks, CA: Corwin Press.

Skovsmose, O. (1994). *Towards a philosophy of critical mathematics education.* Boston: Kluwer.

Skutnabb-Kangas, T. (2000). *Linguistic genocide in education or worldwide diversity and human rights?* Mahwah, NJ: Erlbaum.

Sleeter, C. E. (1997). Mathematics, multicultural education, and professional development. *Journal for Research in Mathematics Education, 28*(6), 680–696.

Solomon, D., Battistich, V., & Hom, A. (1996). Teacher beliefs and practices in schools serving communities that differ in socioeconomic level. *The Journal of Experimental Education, 64,* 327–347.

Solorzano, D. (1998). Critical race theory, race and gender microaggressions, and the experience of Chicana and Chicano scholars. *International Journal of Qualitative Studies in Education, 11*(1), 121–136.

Spencer, J. A. (2009). Identity at the crossroads: Understanding the practices and forces that shape African American success and struggle in mathematics. In D.B. Martin (Ed.) *Mathematics teaching, learning and liberation in the lives of Black children.* (pp. 200–230). New York: Routledge.

Spielman, J. (2001). The family photography project: We will just read what the pictures tell us. *The Reading Teacher, 54*(8), 662–770.

Staples, M. (2005). Integrals and equity. In E. Gutstein & R. Peterson (Eds.), *Rethinking mathematics: Teaching social justice by the numbers* (pp.103–106). Milwaukee, WI: Rethinking Schools.

Stein, M. K., Silver, E. A., & Smith, M. S. (1998). Mathematics reform and teacher development: A community of practice perspective. In J. G. Greeno & S. Goldman (Eds.), *Thinking practices in mathematics and science learning* (pp. 17–52). Mahwah, NJ: Erlbaum.

Stigler, J. W., & Hiebert, J. (1999). *The teaching gap: Best ideas from the world's teachers for improving education in the classroom.* New York: The Free Press.

Stodolsky, S., & Grossman, P. L. (2000). Changing students, changing teaching. *Teachers College Record, 102*(1), 125–172.

Strauss, A. L., & Corbin, J. (1998). *Basics of qualitative research: Grounded theory procedures and techniques* (2nd ed.). London: Sage.

Strauss, A. L., & Corbin, J. (1990). *Basics of qualitative research: Grounded theory procedures and techniques.* Newbury Park, CA: Sage.

Talbert, J. E., & Ennis, M. (1990). *Teacher tracking: Exacerbating inequalities in the high school.* Paper presented at the Annual Meeting of the American Educational Research Association, Boston, MA.

Tate, W. F. (1995). Returning to the root: A culturally relevant approach to mathematics pedagogy. *Theory into Practice, 34* (3), 166–173.

Tate, W. F. (1997). Race, ethnicity, SES, gender, and language proficiency trends in mathematics achievement: An update. *Journal for Research in Mathematics Education, 28* (6), 652–679.

Tate, W. F. (2005). Race, retrenchment, and the reform of school mathematics. In E. Gutstein & B. Peterson (Eds.). *Rethinking mathematics: Teaching social justice by the numbers.* (pp. 31–40). Milwaukee, WI: Rethinking Schools.

Téllez, K. (2004/2005). Preparing teachers for Latino children and youth: Policies and practices. *The High School Journal, 88*(2), 43–54.

Tharp, R., & Gallimore, R. (1991). *Rousing minds to life: Teaching, learning, and schooling in social context.* New York: Cambridge University Press.

Thomas, W. P., & Collier, V. P. (2002). *A national study of school effectiveness for language minority students' long-term academic achievement.* Santa Cruz, CA: Center for Research on

Education, Diversity and Excellence, University of California, Santa Cruz. Retrieved from http://repositories.cdlib.org/crede/finalrpts/1_1_final/
Thompson, A. G. (1992). Teachers' beliefs and conceptions: A synthesis of the research. In D. Grouws (Ed.), *Handbook of research on mathematics teaching and learning* (pp. 127–146). NY: Macmillan.
Turner, E., Celedón-Pattichis, S., & Marshall, M. E. (2008). Cultural and linguistic resources to promote problem solving and mathematical discourse among Hispanic kindergarten students. In R. Kitchen & E. Silver (Eds.), Promoting high participation and success in mathematics by Hispanic students: Examining opportunities and probing promising practices. *TODOS: Mathematics for ALL Monograph, 1,* 19–42.
Turner, E., Celedón-Pattichis, S., Marshall, M. E., & Tennison, A. (2009). "Fíjense amorcitos, les voy a contar una historia": The power of story to support solving and discussing mathematical problems with Latino/a kindergarten students. In D. Y. White & J. S. Spitzer (Eds.), *Mathematics for every student: Responding to diversity, grades Pre-K-5* (pp. 23–41). Reston, VA: National Council of Teachers of Mathematics.
Valdés, G. (2004). Between support and marginalisation: The development of academic language in linguistic minority children. *Bilingual Education and Bilingualism, 7*(2&3), 102–132.
Van Es, E. A., & Sherin, M. G. (2008). Mathematics teachers' "learning to notice" in the context of a video club. *Teaching and Teacher Education, 24*(2), 244–276.
Villegas, A. M. (1993). *Restructuring teacher education for diversity: The innovative curriculum.* Paper presented at the Annual Meeting of the American Education Research Association, Atlanta, GA.
Villenas, S., & Moreno, M. (2001). To valerse por si misma between race, capitalism and patriarchy: Latina mother-daughter pedagogies in North Carolina. *International Journal of Qualitative Studies in Education, 14*(5), 671–687.
Vygotsky, V. (1978). *Mind in society: The development of higher mental processes.* Cambridge, MA: Harvard University Press.
Wager, A. (2008). *Developing equitable mathematics pedagogy.* Unpublished doctoral dissertation, University of Wisconsin, Madison.
Wang, P., Hawk, W. B., & Tenopir, C. (2000). Users' interaction with world wide web resources: An exploratory study using a holistic approach. *Information Processing and Management, 36,* 229–251.
Weiner, L. (2000). Research in the 90's: Implications for urban teacher preparation. *Review of Educational Research, 70*(3), 369–406.
Weissglass, J. (1999). No compromise on equity in mathematics education: Developing an infrastructure. Retrieved on May 20, 2005 from http://ncee.education.ucsb.edu/articlesonline.htm
Wenger, E. (1999). *Communities of practice: Learning, meaning, and identity.* Cambridge, England: Cambridge University Press.
Wiedeman, C. R. (2002). Teacher preparation, social justice, equity: A review of the literature. *Equity and Excellence in Education, 35*(3), 200–211.
Williamson, J., Ranyard, R., & Cuthbert, L. (2000). A conversation-based process tracing method for use with naturalistic decisions: An evaluation study. *British Journal of Psychology 91,* 203–221.
Wilson, S. M., & Berne, J. (1999). Teacher learning and the acquisition of professional knowledge: An examination of research on contemporary professional development. *Review of Research in Education 24,* 173–209.
Wood, D., Kaiser, W., & Abramms, B. (2001). *Seeing through maps: many ways to see the world* Amherst, MA: ODT.

Woods, D. D. (1993). Process tracing methods for the study of cognition outside of the experimental psychology laboratory. In G. A. Klein, J. Orasanu, R. Calderwood, & C. E. Zsambok (Eds.), *Decision making in action: Models and methods.* (pp. 228–251).Norwood, NJ: Ablex.

Yackel, J. & Cobb, P. (1996). Socio mathematical norms, argumentation and autonomy in mathematics. *Journal for Research in Mathematics Education, 27*(4), 458– 477.

Zeichner, K., & Gore, J. (1990). Teacher socialization. In W. R. Houston (Ed.) *Handbook for research on teacher education* (pp. 329–348). New York: Macmillan.

Zeichner, K., & Hoeft, K. (1996). Teacher socialization for cultural diversity. In J. Sikula, T. Buttery & E. Guyton (Eds.), *Handbook of research on teacher education* (2nd ed., pp. 525–547). New York: Macmillan.

NOTES ON CONTRIBUTORS

Vanessa R. Pitts Bannister is an Assistant Professor of mathematics education at Virginia Tech located in Blacksburg, VA. Her research interests include teacher and student knowledge in the areas of algebra and rational numbers, teachers' pedagogical and content knowledge with respect to curriculum materials, and equity and diversity issues in mathematics education.

Tonya Gau Bartell is an Assistant Professor in the School of Education at the University of Delaware. Her research agenda involves preparing and understanding effective teacher professional and preservice development that integrates issues of equity and issues of mathematics teaching and learning.

Dan Battey is an Assistant Professor of mathematics education at Rutgers University. His research focuses on the intersection of elementary mathematics, teacher learning, and equity issues in urban schools as his work supports opening opportunities for teachers to change their practice.

Kyndall Brown is currently the Director of the UCLA Mathematics Project and a recent graduate of the doctoral program of the UCLA Graduate School of Education and Information Studies. He has 13 years of teaching experience in secondary urban classrooms in Los Angeles Unified School District (LAUSD), and has devoted his efforts to professional development for the past 12 years.

Sylvia Celedón-Pattichis is an Associate Professor in the Department of Language, Literacy, and Sociocultural Studies at the University of New Mexico and Co-PI at the Center for the Mathematics Education of Latinos/as (CEMELA). Her research interests include bilingual education, mathematics education, and teacher education.

Notes on Contributors

Angela Chan is a doctoral candidate in the Graduate School of Education & Information Studies at the University of California, Los Angeles. She researches how to improve teacher education to better prepare preservice teachers for the transition to full-time mathematics teaching.

Ann Ryu Edwards is an Assistant Professor in the Center for Mathematics Education at the University of Maryland, College Park. Her research interests include mathematics teacher education and professional development, particularly the use of artifacts of practice as tools for supporting teacher learning about equitable teaching practices, students as learners, and mathematical knowing and learning.

Indigo Esmonde is an Assistant Professor of mathematics education at the Ontario Institute for Studies in Education, University of Toronto. Her research involves an examination of issues of equity and diversity in mathematics education, with a focus on how classroom interactions shape (and are shaped by) student identities.

Mary Q. Foote is an Assistant Professor of mathematics education in the Department of Elementary and Early Childhood Education at Queens College of the City University of New York. Her research interests focus on cultural and community knowledge and practices and how these might inform mathematics teaching practice.

Megan Franke is a Professor in the Graduate School of Education & Information Studies at the University of California, Los Angeles. Her work focuses on understanding and supporting teacher learning through professional development, particularly within elementary mathematics in low-performing schools.

Carla D. Hall is a mathematics teacher at the Civicorps Corpsmember Academy (a charter high school) located in Oakland, California.

Victoria M. Hand is an Assistant Professor of mathematics education at the University of Colorado at Boulder. She is concerned with the interplay of culture, race, power, resistance, and learning in elementary and middle school mathematics classrooms in perpetuating what she calls the participation gap.

Kristine M. Ho recently received her doctorate in mathematics education from the University of California, Los Angeles. Her work focuses on the intersection of teacher development, race, and mathematics. She began her work in education as a secondary math teacher in South

Los Angeles and is currently engaged in professional development work in several schools in Los Angeles.

Carolee Koehn earned her PhD from the University of California, Los Angeles. She currently serves as the Associate Director of the UCLA Mathematics Project and is a former high school mathematics teacher. Drawing upon her experience in instructional support and professional development, she designed and facilitated the Math is More Than Numbers Institute.

Gina J. Mariano earned her PhD in Curriculum and Instruction with a focus in Educational Psychology at Virginia Tech in 2008. She currently works as a Postdoctoral Research Fellow at Behavioral Research and Teaching at the University of Oregon, where she assists with content development for alternative assessments for students with persistent learning disabilities and develops tools to aid teachers in making accommodation decisions.

Mary E. Marshall earned her PhD from the Department of Language, Literacy, and Sociocultural Studies at the University of New Mexico. She was a fellow for the Center for the Mathematics Education of Latinos/as (CEMELA). Her dissertation work focuses on how second grade students communicate their mathematical thinking when they have had access to CGI problem-solving activities since kindergarten.

Sandra I. Musanti is a visiting professor at the University of San Martín, Buenos Aires, Argentina, and a consultant for the Argentinean Ministry of Education. Her research interests include teacher education, teacher development and collaboration, and bilingual education.

Jaime Park is a Lecturer in the Department of Education at the University of California, Los Angeles. She is a faculty member in UCLA's Teacher Education Program and teaches Secondary Mathematics Methods course. Her research interests include mathematics teaching and learning in urban schools.

Jessica Quindel is a math teacher at Berkeley High School in Berkeley, CA. She is also the Cochair of the Math Department and responsible for planning and implementing effective professional development for math teachers. She is concerned with equity and achievement gaps and how teachers can improve their practice to eliminate gaps in student achievement based on race.

Laurie H. Rubel is an Associate Professor at Brooklyn College of the City University of New York. She teaches middle and high school mathematics teachers who teach in urban schools. Her research interests include probabilistic thinking, teacher education, diversity and equity in mathematics education, and the use of real-world urban contexts in the teaching of mathematics.

Rossella Santagata is an Assistant Professor in the Department of Education at the University of California, Irvine. Her research interests are mathematics teaching and learning in different cultures and teacher learning and professional development.

Joi Spencer is an Assistant Professor in the School of Leadership and Education Sciences at the University of San Diego. Her work examines the intersection of race, identity, and mathematics learning. She recently produced the Study Guide to the *MisEducation of the Negro*, which includes culturally responsive mathematics lessons for middle and high school students.

Anita A. Wager is an Assistant Professor of mathematics education in the Department of Curriculum and Instruction at the University of Wisconsin, Madison. Her research interests include professional development in equity and mathematics.

GPSR Compliance
The European Union's (EU) General Product Safety Regulation (GPSR) is a set of rules that requires consumer products to be safe and our obligations to ensure this.

If you have any concerns about our products, you can contact us on

ProductSafety@springernature.com

In case Publisher is established outside the EU, the EU authorized representative is:

Springer Nature Customer Service Center GmbH
Europaplatz 3
69115 Heidelberg, Germany

www.ingramcontent.com/pod-product-compliance
Lightning Source LLC
LaVergne TN
LVHW021657060526
838200LV00050B/2394